Fundamentals of
ECONOMIC
DEVELOPMENT
FINANCE

To my daughters, Wendy L. Rush, Lori L. Abramson,
and Tamara Giles-Vernick, for their encouragement
and continued support of this endeavor as
they chart their own paths to serve others.

—Susan L. Giles

To my daughters, Pieta Blakely and Brette Blakely,
both of whom are great economists.

—Edward J. Blakely

Fundamentals of
ECONOMIC DEVELOPMENT FINANCE

Susan L. Giles | Edward J. Blakely

Sage Publications
International Educational and Professional Publisher
Thousand Oaks ▪ London ▪ New Delhi

For information:

Sage Publications, Inc.
2455 Teller Road
Thousand Oaks, California 91320
E-mail: order@sagepub.com

Sage Publications Ltd.
6 Bonhill Street
London EC2A 4PU
United Kingdom

Sage Publications India Pvt. Ltd.
M-32 Market
Greater Kailash I
New Delhi 110 048 India

Printed in the United States of America

Library of Congress Cataloging-in-Publication Data

Giles, Susan L.
    Fundamentals of economic development finance / by Susan L. Giles and Edward J. Blakely.
        p.   cm.
    Includes bibliographical references and index.
    ISBN 0-7619-1912-0 (pbk.: acid-free paper)—ISBN 0-7619-1911-2 (cloth: acid-free paper)
    1. Economic development-Finance.   2. Community development—Finance.   I. Blakely, Edward J.   II. Title.
    HD75 .G53 2001
    307.1′4′0681—dc21   00-012752

01   02   03   04   05   06   07   7   6   5   4   3   2   1

Acquiring Editor:	Marquita Flemming
Editorial Assistant:	MaryAnn Vail
Production Editor:	Diana E. Axelsen
Editorial Assistant:	Kathryn Journey
Designer/Typesetter:	Rebecca Evans/Lynn Miyata
Indexer:	Mary Mortensen
Cover Designer:	Jane M. Quaney

Contents

Acknowledgments

After working together for almost 20 years, we have learned that local economic development does not occur in a vacuum. We have watched community leaders, local citizens, entrepreneurs, nonprofit groups, business leaders, and developers become increasingly frustrated with the issues that cannot be solved unless sufficient money is available. Though many scholars and practitioners have struggled with the subject of finance, we too have been often puzzled by the inability of many to use successfully the tools that are common for businesses. Knowing where the money is and how to access it becomes central to any business, whether it is a product or a service.

We are truly indebted to our colleagues and business associates who have urged us to forge a new path by looking at local economic development finance in the context of people, community, and money. As we state in the following pages, the essence of financing local economic development is the act of combining values and perceptions. All parties—the public sector, the private sector, foundations, and nonprofit organizations—can achieve their collective goals because they understand the agendas. If they choose to move forward in a transaction, they bring their total resources together to create wealth for the community, its citizens, the developers, investors, and financial institutions.

We have been supported by many in the process of writing this book. Our students at the University of Southern California have been invaluable. We are especially grateful to Sarah Imber, Matthew Brian, Bryan Payne, Joshua Hart, Susan Kim, Amitabh Barthakur, Efrem Joelson, Glen Maestre, Terri Dickerhoff, Deborah Amshoff, Sevada Hemelian, Karen Hsu, and Bonnie Montoya for their comments and participation in some of our cases. Ellie Tostada provided important administrative assistance as we worked through the manuscript. We thank our business colleagues in the cities of Oakland, San Francisco, Sacramento, San Jose, Los Angeles, San Diego, New York, Boston, and Washington, D.C., and those at Bank of America, Wells Fargo Bank, Salomon Smith Barney, Stone and Youngberg, Merrill Lynch, the Fannie Mae Foundation, and the California Public Employee Retirement System. Our friends and families have also given us the encouragement needed to see this book to completion.

Introduction

"For I know the plans I have for you," declares the Lord, "plans to prosper you and not to harm you, plans to give you hope and a future."

—Jeremiah 29:11

At the beginning of a new century, the role of government in providing social and community needs has changed dramatically. For most of the 20th century, the Great Depression defined government expansion into almost every facet of community life, from child welfare to housing and community facilities. But beginning in the 1970s, with a rising national debt and shrinking public support, along with compassion fatigue for more social programs, the national and, to some extent, even the local government role as the direct provider of community-oriented resources has diminished.

Today, if a community requires a park, a child care center, a new program for drug addicts, or a youth soccer field, an appeal to government is seldom the answer. Low-income neighborhoods especially, with the new wave of economic development and the decline of welfare, must find new resources from within.

Moreover, when the private sector initiates projects ranging from opening new firms to developing a local movie theater, the public anticipates some form of offsetting public amenity or gratuity to meet local needs. As a result, when a factory opens, the new company has to go beyond providing jobs. It is expected to contribute to community good by doing such things as building child care facilities for workers accessible to the community or providing for senior citizens. Today, when a developer takes on the task of resurrecting the local downtown economy or providing needed affordable housing, he or she is expected to provide funding for public benefits. These *linkage programs* require specific project funds from the developer in exchange for the right to develop within the city core. The developer may be required to provide funding for a community arts program, recreation, housing, senior care, a youth orchestra, youth soccer, or other public benefits.

There are some good reasons for the provision of community benefits as a requirement to use public assets or receiving public licenses or assistance. The most compelling rationale for requiring a developer to pay fees or provide an off-setting public benefit is that the person receiving public largesse is using irre-placeable civic assets to increase personal wealth. As a result, the developer is entering into new arrangements with community groups and government agen-cies to develop a new range of public-private goods and services through various forms of development partnerships.

Clearly, community projects and programs cannot await the payment of development fees for special projects. Few communities can simply raise taxes to provide for special programs and services to meet community needs. In many instances, tax rates are restricted by state law (as in the case of California's Propo-sition 13 or Massachusetts's Proposition 2½). Therefore, the community simply does not have the fiscal capacity to fund important or beneficial community pro-grams. Communities both large and small look for unique arrangements to create facilities that are cost-effective and pay their own way via fees or sales of services or both.

Low-income communities are increasingly interested in forging their own economic destiny through community economic development projects, such as community-owned and -operated ventures; community services, such as commu-nity health, education, and nutrition; and community-owned and -operated busi-nesses. These new ventures increase local employment and retain local buying power.

The concept of "new markets" has entered the national lexicon by creating market strategies to replace welfare approaches as the fundamental attack on pov-erty and economic disenfranchisement. As James Carr (1999), vice president of the Federal National Mortgage Association, stated, "Wall Street has a variety of inno-vative investment vehicles that can be tailored to rebuild distressed communities" (p. 2). As a result, for the first time in more than 50 years, the ghetto is viewed as a new opportunity to create wealth.

Finally, many communities, both rich and poor, need to contribute their own time, money, and labor to support local voluntary efforts. Organizations such as the American Red Cross, Boy and Girl Scouts, parochial and private schools, Little League, and local symphonies and museums can no longer be sustained by com-munity giving. These civic activities must use specific methods—renting part of their facilities to other nonrelated tenants or engaging in other businesses—to augment cash flow to continue their respective activities. Public goods such as convention centers and roadway systems are part of a newly privatized market-place. Communities or community groups can no longer wait for private charity or public assistance. Public projects have to incorporate the same kind of financial market competition to meet the test of fiscal worthiness to attract either public or private capital.

In the following chapters, we discuss strategic business planning tools to be used to link public community funding and private marketplace financing. It is important to recognize that local economic development is not necessarily a financially feasible endeavor. Linking community funding with private marketplace financing can solve the feasibility problem for public agencies, developers, and lenders. We describe the need for community funding and financing strategies that originate from the organizational and institutional forms of community projects.

The strategies to close the financial gap become visible through the business planning process. This process readily identifies a project lacking short- and long-term financial stability. The process can provide mechanisms to link public funding with private funds so that cash flow will be available to service debt and provide reasonable returns to investors. We show how private development can incorporate community programs as an asset to the development project or programs, thereby increasing the opportunity for public funding.

The objective of this book is to provide the necessary tools for "hope and a future." We illustrate how financial tools can be used by non-financially skilled planning and economic development professionals, policymakers, and students. We outline the critical elements of the strategic business plan and financial processes so that local economic development projects can secure the funding for development. These processes will enable a community to build its future.

Planning, economic development, and investment professionals have typically used feasibility studies, market analyses, and discounted cash flows as the methods for evaluating projects or "deals." Today, projects are more complex and competitive, require varying degrees of financial dollars from the public and private sectors, and they entail enormous risks for cities if the projects do not succeed. For that reason, a more detailed strategic business planning approach is needed to eliminate the extremes in forecasting the high yields or the downside exposures.

Our thesis is the following: To finance local economic development, the strategic business plan is the preferred vehicle. The business plan describes the organization and the project in real terms as measured by the financial community. It is a rational approach to business decision making. It enables the local economic development project to be competitive with other projects seeking funding by investors and institutions. The strategic business plan enables the developer, the public sector professional, and the financial partner to understand all facets of the urban enterprise business so that investment will occur.

The components of the urban enterprise as presented in the business plan approach are (a) the strategic intent, (b) the organization and management, (c) the project, (d) the market, (e) the marketing and sales strategy, (f) the financial analysis, (g) the benefits and risks for the investors, and (h) the implementation program. The extent to which we can examine and review the components will minimize risk and lead all parties in the transaction to develop rational decisions for investment.

As we work our way through the business plan approach for economic development finance, we will review the four major steps of conducting a business analysis and how each leads to the next.

The synergy between the market for the project and the investment in the project begins at the basic level of understanding what development is required versus what development is desired. Finding the market is the key to success. The opposite is also true—not knowing the market is the key to failure. The *market for building* includes both cases and exercises on demographic analysis, real estate market research, and locational analysis. As we refine the market analysis, we will move to the *fundamentals of local economic development finance,* examining both the private and public sector approaches to decision making. Understanding the money sources and how to secure gap financing is a requirement for successful and stable economic development projects. Our case studies and exercises will enable the reader to use a "hands-on approach" to financial analysis, building from assumptions on market analysis through the layering of various financial instruments.

Once the project is refined and assumptions for market and finance are adjusted, we examine the organization structure best suited for implementation of the project. How much money do we really need? What elements are necessary to build an organization to ensure the project is approved, developed, financed, marketed, leased, or sold? Who are the members of the organization? Is this a public/private partnership? On the subject of financing the organization, we examine small business issues and funding sources. From there, we develop a small business financial plan, focusing on start-up expenses, debt and equity, and the operating pro forma. Readers will be asked to work through the exercises for building a "start-up" organization.

We develop the link between public community funding and private marketplace financing. Increasingly, the distinctions between the two forms of finance are diminishing. We describe the need for community funding and financing strategies, the organizational and institutional forms for community projects, and the fund-raising strategies or approaches for community/social projects. We show how private development can incorporate community programs as an asset to the development project or programs. Finally, we describe how community groups can incorporate private economic activities into their nonprofit activities. The book's appendices offer a sample business plan (Appendix A) and a glossary of terms (Appendix B).

This book is about organization and about money—how to get it and how to use it. The components of the business planning process work as a financial plan for the lenders and investors and a management tool for the organization. The emphasis throughout the book is to understand the agendas of the developer, the city or county, and the lenders and investors.

As many of us have observed, public sector orientations to planning, community and social development, or local economic development often overlook the money end of the project and focus almost exclusively on the social good. Unfortunately, plans do not materialize without money.

The goal of this book is to provide the reader with the tools to acquire the money that is necessary for any project to work. The book is intended as a follow-up to Blakely's *Planning Economic Development: Theory and Practice* (1994, 2001).

* * *

This book is not a substitute for the more comprehensive real estate works and/or business finance. Rather, it is intended to outline the critical elements of financial planning for local economic development that include financial and organizational components.

Local Economic Development Finance Design

Local economic development is the field of study and practice in which the theory for this book is placed. In local economic development, the aim is to improve the social, employment, and physical conditions of a community or area (Blakely, 1994, 2001). In this respect, local economic development differs from other forms of economic development that focus on the national, regional, or international scale. These more macro programs are aimed at money supply, trade, and competitiveness. Local economic development is oriented toward micro changes in a specific area.

In local economic development, our focus is on the micro issues of firm starts and community quality of life, together with the institutional issues of government actions and interventions. The goal is to raise the quality of life for communities and disadvantaged residents and move communities into the mainstream of the national economy. Community residents, consumers of services, local government officials, lenders, and investors are concerned about mechanisms to create a self-sufficient economy that can sustain the residents of an area, improving their lives and increasing the opportunities for them to achieve economic wealth and stable family living.

The premise here is that local communities in a capitalist economy must attract capital and create markets to be revitalized. These in turn generate more and better jobs to improve the housing and business atmosphere in these localities. This book acknowledges the current options regarding opportunities for development in disadvantaged communities. One can take the perspective of Michael Porter (1995), who suggested that most low-income areas in the United States are ideally located to rebound and develop internal mechanisms for their own development by competing in the regional market and better satisfying local demands for goods and services such as food, banking, and other retail needs as well as regional transportation. Blakely (1994, 2001) and others (Boston & Ross, 1997, pp. 161-162) suggested that the conditions of local disadvantaged communities in the United States go well beyond the market failures that Porter described and argued that there is a need to generate community social capital by microenterprise and similar strategies that build new economic and social resources. Many mechanisms have been devised to do this, ranging from area development strategies such as enterprise and empowerment zones to the stimulation of new local businesses or enterprises with microloans and other vehicles to encourage local ownership and employment. None has taken the position that local economic development must compete in the capital markets with conventional projects.

It is important to push the boundaries of local development to include creative and innovative projects that serve as catalysts for larger and more comprehensive development. The repositioning of the area's economy through the development of new manufacturing, electronic infrastructure, or a community of dot-com companies should not be overlooked in the planning and financing of local economic development. These catalyst projects will use conventional and unconventional sources of funding to ensure that local development occurs.

There are several thousand local governments that engage in some form of local economic development planning and financing. Local governments have a variety of financing vehicles at their command, ranging from the use of tax receipts to the borrowing of funds on the municipal bond market for infrastructure and redevelopment. Local governments most frequently use these funds for physical improvements such as street resurfacing, housing, and parking structures.

Public sector tools for planning, community and social development, or local economic development often overlook the fiscal worthiness of a project in order to secure the desired economic development opportunity. Plans must be strategically developed, tested, retested, and presented in a clear financial framework to secure both the public and private dollars to implement the plans.

In this chapter, we will work through the strategic business plan approach and review four steps for conducting a development finance analysis, showing how each step is a prerequisite for the next. In the cases and exercises, we ask readers to identify an economic development finance strategy and review the types of organizations that can deliver the product. Readers will be asked to work through exercises for building a "start-up" organization or working within an existing organization.

The Context for Local Economic Financial Need

Local community economic development needs to be rooted in the history and the human settlement pattern of an area. The area can be as large as a city or as small as a neighborhood within the city. In most instances, the area of need or economic revitalization is depressed, with high unemployment or deteriorating buildings and physical infrastructure. The need might be for a new concert hall for the city as a whole.

No matter what is needed, the first order of business for the development specialist is to know and understand the history of the area and the politics of the community. In addition, the specialist must know locational constraints, past market performance, and present conditions, as well as forecasting the possible development options or needs. In most cases, the entity developing or designing the development strategy is a local government or nonprofit or a joint venture entity that may be a blend of nonprofit and government, such as a local development corporation. In almost all cases, these development organizations will work with or collaborate with profit-making partners.

There are millions of nonprofit voluntary organizations and community groups that work for the civic welfare. The needs of the nonprofit organizations, community groups, economic groups, businesses, and city governments compete for money available from nonprofits, foundations, federal/state/local governments, and private financial institutions. Developments and services must be paid for, regardless of whether they are a public stadium, retail mall, golf club, or community center. Volunteer labor or the donations of equipment and other services will not be sufficient to cover the cost of development or provision of services for a community. Whether the organization is a small community group or a large city, if money must be raised or generated for the community good, the organization or government entity must be prepared to compete in the private financial market.

The Capital Gap in Financing Nonprofits and Nontraditional Institutions

Access to capital markets is an increasingly difficult problem for nonprofit and nontraditional local finance organizations. The size of the nonprofit community capital market is growing as a result of the increasing need for banks and other institutions, including foundations, to find mechanisms to invest in low-income neighborhoods. For banks, the Community Reinvestment Act (CRA) sets the standard and establishes a need for finding appropriate investments in low-income communities.

New vehicles for such investments have been created as community development financial institutions (CDFIs), authorized by the Comptroller of the Currency, act to channel funds back into local neighborhoods via subsidized or government-backed lending. There are now many such institutions in the United States. CDFIs look and act very much like traditional banks, except that they do not have a retail operation. They are making the risky subprime loans to small minority start-ups and other community investments.

In addition to the CDFIs, many cities have devised their own local fund mechanisms to stimulate new business formation through revolving loan funds (RLFs). RLFs provide the very small loans for local vendors, hair salons, and other community-serving enterprises in areas of the city that seldom receive lenders' attention and where the loans are so small that the transaction costs are well below a banker's thresholds.

Finally, foundations at the local, national, and international level are lending funds for social purposes through their investment portfolios, expecting returns on their investments at below-market rates. In a few instances, social funds traded on the capital markets have been established to invest in both national and international projects aimed at selective improvement in the environment through investments in forestry, farming, and even small business microloans for farmers and artisans.

In essence, the community capital market is taking on a new character. It is not just local housing and real estate development but broader areas of asset formation in business and community services. There is a "skills and information gap" for those involved in local and community development financing. This gap prevents professionals and practitioners from navigating through the unfamiliar territory of attracting market-based capital by meeting the strenuous requirements of project financial design and reporting. The following chapters provide the tools to enable these professionals and their projects to be competitive for the financial resources that are available.

Strategic Development Finance Design

Capital markets, large or small, require that the justification for funds be organized into a plan. In the case of local development planning, a strategic business plan or development finance plan provides the set of information, analysis, organization, and financial parameters that can be presented to outside investors. The strategic business plan provides the investors with the information they need to know about the venture and how much they can expect to make for each dollar invested in the project.

A good strategic investment plan includes the following: (a) the goal or project intent, (b) the organization and management approach, (c) the venture selection process, (d) a detailed project description, (e) a clearly defined market for the project, (f) the marketing and sales strategy, (g) a well-developed financial analysis, (h) an assessment of benefits and risks for the investors and consumers, and (i) an implementation program. The extent to which one can examine and review the components will minimize risk and lead all parties in the transaction to develop rational decisions for investment (see Figure 1.1).

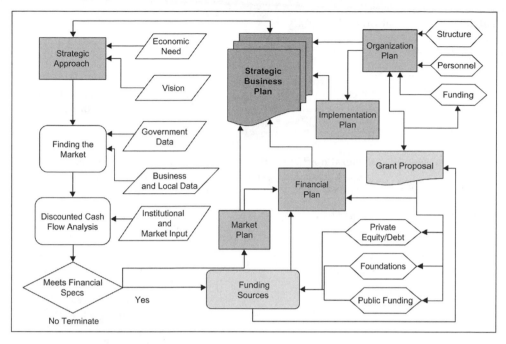

Figure 1.1. Strategic Business Planning Process

Local Economic Financial Analysis

The business plan should be based on four types of analysis: analysis of the local economy, market analysis, feasibility analysis, and investment risk analysis (Table 1.1).

Analysis of the local economy is central to identifying demand for a particular project or service. Consumption patterns, spatial requirements, business development, and services have a synergistic relationship so that new business formation can occur. For example, developing an incubator technology center or a low-income housing project in a local community will not just require the availability of land. A technology center must be located close to the skilled workers in a community, providing incentives for development and services to support the new industry and businesses being developed. For the low-income housing project, a thorough understanding of the population and income composition is essential. A further knowledge of the community incentives and services to support the housing population and physical plant is also important. No matter how the local economy is performing, some group will be underserved or not well served by the current situation. Therefore, an analysis and understanding of the local economy is central to identifying demand and need for a particular development.

The availability of money to finance the project will be predicated, in part, on the detailed documentation of the local demand as well as the broader market

Table 1.1 Types of Business Analysis

Type of Analysis	Focus	Functions
Analysis of the local economy	Overall consumption capacity, growth of the geoeconomic area, and demand	To identify users or demand for the goods or service geographically
Market analysis	General market for the project, program, service, organization	To assess the size of the market, the competition for a venture, and the ability of the venture to compete
Feasibility analysis	Project, program, service, organization	To evaluate financial costs and benefits of the project
Investment and risk analysis	Investors'/lenders' decision	To assess venture/project, program, service, and/or organization as a potential investment for repayment or profits; to evaluate investors' exposure to loss of investment or loan

demand. Thus, the first step in conducting the business analysis must emphasize knowledge of the determinants of demand—what elements will make the project work.

Market analysis, the second step of the business analysis, focuses on the general market for the project, program, service, or organization. The goal is to outline the supply of and demand for particular goods and services. Demand is documented by a clear, well-presented case of need for the project; supply is created by investment related to the amount of competition there is for the goods or services; and investment is the capital or money made available for which some return is required to make the project feasible.

Building on the knowledge of the local economy, market analysis focuses on the components of business, trade, and consumption by differing population groups. Market analysis isolates the supply and competition—existing in the local economy *and* planned for the economy. Market analysis addresses what the users want, what will compete with the project, and what the market can afford to pay. For our example of the technology center, an examination of the existing local businesses and local technology centers is critical. For our example of the housing project, a review of similar local projects for type, size, and price is key. Balancing supply against demand to determine the market niche is another aspect of market analysis. We will discuss market analysis in greater detail in Chapter 4.

The *feasibility analysis*—an evaluation of the nature of the venture or project based on financial costs and benefits—is the third step of the business analysis. A feasibility analysis addresses the financial costs for the project and measures the benefits that accrue to the investors and the community. The financial analysis incorporates market assumptions and financing assumptions to arrive at an investment return that is compared and contrasted with similar investments and alternative investments. The benefits to the community are measured in a framework that addresses jobs, risk, spending patterns, and tax revenues.

Finally, the *investment and risk analysis* is the most crucial step of the strategic business plan. Indeed, it is the reason for the business plan. Without investment, there is no demand or development. The purpose of the investment analysis is to assess the project, program, service, or organization as a potential investment, given all other alternatives. The analysis clearly evaluates the risks and returns associated with the project.

Investment and return requirements, short- and long-term holding period, tax benefits for the portion of the project where tax benefits can be used, and investment and market risks are but a few of the measures examined. Given two opportunities, would an investor be better off financially placing money in a technology center or a low-income housing project? The answer depends on the investor's risk profile, cost of money, and return requirements.

Conducting the four steps of business analysis enables us to focus on the core issues central to the business plan and reach the necessary conclusions regarding financing. More specifically, the steps of conducting the analyses lead us through an assessment that will enable us to isolate the money source—the entrepreneur investor, the institutional investor, the venture capitalist, the traditional lender, a public agency, and/or a foundation—needed for the project, program, service, or organization.

Strategy is always required before feasibility. We need to know who we are and what we want to do before embarking on any approach to measuring the market or assessing the value of the program or service. The strategy and development goals are aimed at solving quality-of-life problems within a community. The goals include job creation and increased revenues for services and additional investment within the community or urban center. New ventures are required to create jobs, increase local income, and bring social stability to the community.

Developing the Strategy

Strategic planning for community and social projects incorporates a broad set of development approaches (described in Blakely, 1994, 2001) that include many different activities, projects, and entrepreneurial ventures. In one way or another, all ventures have as the ultimate goal creating and increasing the economic resources of a community so that residents can control or increase the quality of life with a minimum of dependence on outside help. Not all potential activities meet these standards or generate the requisite resources, such as money, skills, ownership, and political influence. Furthermore, not all projects create or increase the community's quality-of-life factors to the same extent.

All community groups require investment capital, skills, and political power to be successful. But not all organizations and groups require these resources in the same ways, amounts, or combinations to operate successfully. So individuals, government agencies, and groups interested in community and social programs are faced with an important set of decisions. They must decide what courses of action to take given the resources they possess or can obtain.

Moreover, the group or agency must decide which activities or projects are best at a given time and place. They must address the following questions:

1. What are the general objectives?
2. What are the priorities of the organization or agency?
3. What trade-offs or compromises will be made when funding for the program, development, or service is secured?

In developing the strategy, the group must decide whether their priority objectives can be realized soon enough or completely enough to justify the time, energy, and money committed to a project. They must consider whether a project using one approach will reduce the chances of reaching other objectives. For example, if a community raises money for a low-income housing project, it may not be able to raise the funds in the same year for a day care center and playground for the new technology center.

The most fundamental element of a strategy is to clarify the vision of the organization and its mission. The group proposing the activity must identify the vision of the organization and how the intended project matches the vision. The next decision is to determine how the venture will work and who will directly benefit from the venture. Finally, the group must examine whether it will use internal financial resources for the project. The decision-making process will clarify the values of the group and help it focus on external resources required to complete the project.

Most community ventures are designed to involve and benefit either a large group of people or a few. Cooperatives (co-ops) and credit unions are examples of economic development businesses that are community based and have shared ownership as a primary objective. When financially successful, co-ops typically do not provide substantial income for their owners because the returns are small and must be divided among many owners. On the other hand, a small, privately owned business, if successful, may provide a significant increase in income for its owners, although ownership is usually limited to a few.

What is the best organizational approach? Is it better to share ownership widely with little immediate financial gain for those who are owners, or is it better to help a few achieve financial gains, with no guarantee that their success will make any difference for others in the community? If it were possible to do both and even more in every community where economic development was needed, clearly that would be desirable. However, this rarely occurs. Given that community projects require the very resources they create—people and time—both are in short supply in communities of all income levels.

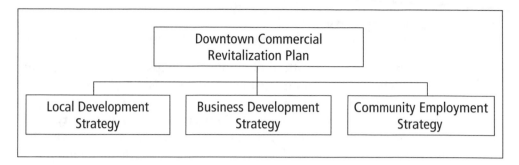

Figure 1.2. Inglewood Revitalization Plan

Choices have to be made about the best use of all community resources. These strategic decisions will form the value proposition for the organization or group. In some cases, value choices may not be available. For example, in communities in which low-income residents are split into many competing groups, little can be done. But in cases where choices are possible, decisions have to be made. One of the key decisions is who gets what and how much. Put another way, is it more important to invest resources in activities for community gains or for individual gains? In some ways, the answer to that question seems obvious. If community economic development is the strategy for trying to solve community problems, then community gains are primary.

Spending patterns resulting from successful businesses generate multiplier effects throughout the community. For example, if the business is financially successful, the owner's income is increased. If the owner uses some of the increased income to buy personally needed products and services or business supplies from other businesses in the community and the region, he or she helps contribute to the success of those businesses as well and so helps raise the income of their owners.

Establishing objectives and priorities within the context of the strategy serves to narrow the range of alternative activities that any individual or group may consider. The availability or lack of resources for specific activities will also serve to narrow the alternatives possible.

Understanding the economic climate of the target area or local economy will enable the organization or community group to identify the appropriate project or service that will succeed within an urban area. Community groups and local government agencies can take advantage of new opportunities by developing and maintaining close contact with local businesses. *In economically depressed communities, the question will not be where to look to find a local economic development project but how to best find and use local business, the community's limited resources, and conventional and nonconventional sources of funding to meet the revitalization needs of the area.*

A downtown commercial revitalization plan may seek to integrate the strategies of the locality, the community employment requirements, and the greater business and financial community, as illustrated in Figure 1.2. This venture depended on group cooperation.

Organizational Capacity

Many community-based organizations or nonprofit groups envision their becoming a conduit for investment coming into the community. Local government agencies similarly view their own regulatory and funding powers as the key decision point for investment into the community. Both are correct; community-based organizations and local government agencies often work in tandem to accomplish broad community strategies and objectives. Community or local development corporations (CDCs or LDCs) act as a source of investment capital through their access to government and special foundation funding. CDCs are engaging in business development activities in minority communities all over the nation. However, new CDCs or other venture-oriented nonprofits must recognize the strengths and limitations of this organizational form before commencing the risky process of starting any service or business.

Expertise

Involvement in a new business venture requires a sound strategy, a business plan, and the internal organizational expertise to execute the plan. If a business venture is proposed that will actually make money or engage in transactions, the board of the organization will need to establish its own credentials or partner with a business development arm that is properly staffed to organize and conduct the business of providing development or services. Community groups and CDCs can identify these resources from the outside, initially forming a business arm or committee that includes representatives from the business and financial worlds who have experience and skills in the development and/or management of business and a commitment to the community development processes. Typically, board members and/or CDC members and community beneficiaries are included so that they can become educated in business activities. The group can make recommendations to the board on venture selection and feasibility and can even provide training to those without venture expertise. It can partner with more qualified groups to seek interim, short-term, and long-term financing for the project.

Track Record

All organizations, whether private entities or local community groups, must demonstrate their capabilities through completed projects. Financial institutions, foundations, or grant organizations will be more inclined to risk capital when an organization has an established track record. Community groups must demonstrate that they can execute any new project or program. If the community group is considering financial involvement with an organization, the partner or investor entities will seriously examine the proposed capabilities of the group. It is important that the community group be able to prove that it is a stable organization, capable of managing funds and making informed decisions on behalf of the community. The use of well-recognized accounting and managerial procedures shows the expertise and viability of the organization. An organizational brochure that highlights previous projects and the capacity of the organization to mount large-scale efforts is important here. Further, a demonstrated knowledge of the economic conditions and resources of the target area as well as a history of well-planned and completed projects will add to the strength of the group. For new CDCs or community-based organizations, a series of small projects, effectively handled, will begin the track record and pave the way for larger and more complex ventures. The same is also true for private entrepreneurial development companies.

Form of Ownership

Any project undertaken will create ownership options and ultimately financial options. In some instances, the project is owned and operated by the investors, the developer, the community group, the public agency, or local individuals. In other cases, the organization has the capacity to successfully operate the venture alone. In other situations, the project may become a "joint venture" where several businesses share the ownership, one group manages and operates, and others are silent partners. Joint ventures and partnerships are created typically to benefit the project and the community and reduce the risk elements of the venture. The issue here is which option makes the most sense to secure the development rights for the project and obtain the required financing to implement the strategy. A demonstrated track record is important. If the group has no track record, then some form of joint venture will be required to compete for the development rights and the financing.

Using the project assessment matrix in Table 1.2, we can compare and analyze multiple projects to determine opportunities for development and begin developing the development strategy.

Resources

The degree to which the community can mobilize financial resources behind the project is an important indicator as to whether a project can be developed and implemented. Some form of direct equity is required. Foundations and other investors have very strict requirements as to how much local "hard money" must

Table 1.2 Project Assessment Matrix

Criteria	Project 1	Project 2	Project 3
Organizational mission and capacity	To create new employment	To increase wealth by expanding technology base	To develop community arts facility and program
Community needs and market	Manufacturing facilities—no facilities close to community	To assemble land for tech park; to establish small tech facilities in neighboring communities	Downtown land assemblage and program; no competition in adjacent towns
Resources and expertise	Limited capital and limited development expertise	Economic development funds; job training funds; tax incentives from state; strong expertise	Economic development funds; grants from the National Endowment for the Arts; other local funds; limited expertise
Track record	No development track record	Developed tech parks in three states	None
Form of ownership	Joint venture	Public-private partnership	Partner with experienced group

be invested in a project. A developer or community group might raise its initial capital from a variety of local sources, including churches and local fund-raisers. Clearly, if little or no local money will support the project, very few outsiders will risk their money unless the community can demonstrate that "in-kind" contributions will be made or some outside resources can be obtained.

All economic development projects require money to start and additional resources to maintain and operate them. Resources for economic development begin with the community and take the form of labor, time, "sweat equity," political savvy, and networking. As a result, economic development is now increasingly the language that businesspeople understand. It requires resources and skill building within the community. It involves seeking assistance and cooperation, which may directly benefit the larger community. Getting the city director of recreation, for example, to provide recreation facilities in a low-income neighborhood may often be impossible if the director argues that the budget is not large enough. It may be much more possible to purchase recreation equipment from a minority-owned cooperative in the community for a small community-owned day care center.

As economic development becomes more popular as a sound approach to dealing with poverty, the chances increase that businesspeople, bankers, lawyers, accountants, big corporations, and governments will make themselves and their

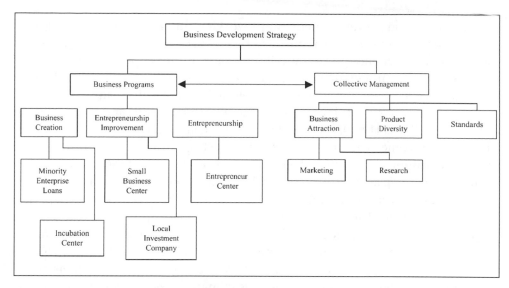

Figure 1.3. Inglewood Business Strategy

resources available to low-income communities and the redevelopment of urban centers. The need for these external resources and skills is great, but the community has an important concern: It wants to maintain control rather than pass the economic development activities and control to outsiders.

In the case of the downtown plan shown in Figure 1.2, the organizational strategy took the form shown in Figure 1.3. The resources of the building owner were used to renovate two upper floors for small business development. Though the building owner's mission was to provide educational opportunities for high school graduates, the owner perceived the need for small business development within the community. The collective efforts of management, the city, and the community resulted in the development and financing of the business incubator center.

A community group or business entrepreneur might need to raise money from local businesspeople to start a business activity. In some cases, the local businesses might participate by agreeing to reserve seats on the board of directors for business. The businesspeople could be an asset by providing both funds and business expertise. However, depending on the number of board seats provided to the local businesses, they could easily be in a position to determine the activities of the corporation beyond a point acceptable to its owners. If one of the primary objectives of the corporation is to keep control in the hands of the owners, the owners may need to limit the number of board seats that are available for the financial/local business partners.

Just as the cooperative attitudes of key people outside the community may be a resource in economic development, the leadership skills of residents within the community are vital resources. Mobilizing resources for economic development activities both within and outside the community depends on the presence of strong leaders and strong, unified groups working with the leaders. A simple grocery cooperative or other local business cannot succeed if it cannot get enough members to join or enough community business. If the cooperative members or residents do not participate in its activities, if they shop for most of their food or other goods elsewhere, the cooperative or local business will collapse.

Low-income residents buy food and purchase other goods and services. Some authors, like Porter (1995), consider low-income communities new repositories of disposable income. Therefore, one of the most important tactics of any low-income community is retaining the money it spends outside of the community for basic services such as food and shelter. No matter how much is available within the community, very little of it is available for investments in new enterprises. Though it is difficult to do, in some cases the initial capital for a community project can be raised from local established institutions such as churches, schools, local credit unions, and community fund-raisers. In other cases, local business-people will support worthwhile community projects that advance their enterprises or better the community's business climate. None of these techniques is likely to raise nearly enough capital to initiate a project of any scale.

Local or community fund-raising efforts are very important. They help focus community attention on the project and mobilize necessary political support. In some instances, it may never be necessary to use the community money at all. Local corporations may be sufficiently impressed by the community effort to take over the project and use their resources of both money and personnel to complete it. But unless the community involved makes some investment of its own, it will have little control over the decisions made, no matter how much money it raises or how much support it mobilizes.

Another important resource in economic development is the knowledge and attitudes of local residents. A community that is vital with cohesive purposeful leadership is well on the way to success. Understanding community attitudes and communicating this knowledge shapes the responses of local government to local initiatives. Attitudes are more important than money. If the community has reached consensus on its vision and strategy, it can attract money and power.

Finally, every community has historic resources and other cultural resources that need to be documented and added to the community resource base. Boarded buildings and other dilapidated structures may prove to be resources if they can be reused creatively as loft apartments or new software facilities. The community must organize and focus its resources before it can expect investment from any outside investors.

A community strategy or direction for the overall economic development of the area is the first order of business. After an overall strategy has been developed, individual projects follow.

Case Study 1 provides an illustration of the range of decisions any organization has to make to determine how to configure itself to embark on financing a project.

Case Study 1: Brava Valley Community Corporation

Faced with a tightened budget and diminishing revenue from the federal government, the Brava Valley Community Corporation (BVCC), serving seven counties in
central New Mexico, began looking into economic development as an alternative
to grants and subsidies. The agency wanted to establish the groundwork for long-
term improvement in the conditions of the poor and to increase the self-suffi-
ciency of both the agency and those it served.

Initially, BVCC focused on the lack of infrastructure such as roads, sewers, and
transportation systems. For these efforts, the BVCC was able to issue tax-exempt
revenue bonds as a community service district. The community received a B rating
on its bonds and has made all of its debt service payments. After helping several
communities obtain water and sewage systems and low-income rental housing,
the agency began to correct the lack of effective transportation services both
within the towns and linking the widely dispersed valley population to jobs and
services. In an area where more than a third of the 150,000 people fall below the
poverty line, many local people who otherwise could have gotten jobs could not
afford transportation to work. Also, the agency realized that "it was no good to have
a family planning clinic if no one could get to it."

Now, 5 years later, most of the programs BVCC is running include a small transit
component. Each division has its own car or van under the auspices of a program
director. Currently, BVCC has 25 minibuses and vans; two one-story buildings, each
with 10,000 square feet; and an old downtown single-story store with 30,000
square feet in Brava City that was given to the corporation by JC Penney as a charitable gift. The building cannot be sold.

BVCC has a staff of more than 100 employees. It operates the local senior Meals
on Wheels program for the countywide elderly program as well as 16 low-income
apartment buildings with 550 total units. BVCC is debt-free except for a few self-
amortizing housing loans totaling $500,000. The BVCC board includes a banker, a
lawyer, the local chief of the Nampo Nation, and several community leaders, including one member of the city council.

The board is divided over whether it should be reaching more of the Native
American population with serious employment problems or serving the broader
community with economic development programs that improve the valley's overall unemployment rate of more than 15% for the prime employment age groups.
The board and staff are at odds over what to do about the pending closure of the
Tailwind Air Force Base. The board thinks the base would be a major asset for a new
industrial center and could cure the chronic housing problems for all segments of
the community. The base was an air maintenance facility and contains some of the
best hangars and maintenance equipment in the nation. On the other hand, the
staff feels that the process of military base conversion is well beyond local capabilities and could be an endless swamp of details that could prevent the BVCC from

using the skills they have for family planning and other services to meet current community needs.

BVCC Executive Director Henry Mohamid-Gonzales has asked Brava Valley College to help look at other ventures that BVCC might take on to diversify its financial base. As a result, a Brava Valley College class in economic development finance has been asked to develop a set of new ventures that might serve the community and fit with the goals of the BVCC board.

Case Exercise

Assume that you are a consultant asked to assist BVCC. You and members of the class have been asked to make a presentation to the BVCC board on how to restructure the goals and investment opportunities of BVCC. You have been asked to demonstrate how your group will conduct an analysis of needs and help the board and staff select a venture. Develop a short bullet-point presentation for the board meeting. You may use PowerPoint or overheads for the presentation.

The Strategic Business Planning Process for Local Economic Development

The need for money is common to all businesses or services, whether they are government organizations, service businesses, nonprofit organizations, or private businesses. In any project, the need to estimate not only the cost of facilities but the ongoing costs of doing business is an important aspect in determining whether the business venture is viable.

Entrepreneurs, developers, and service providers are increasingly called upon to state their viability as business enterprises before securing capital for a new technology project, a real estate development, or a community job training program. Financial supporters, joint venture partners, traditional lenders, public agencies, foundations, and venture capitalists desire business plans before committing to investments. Business plans, therefore, become an integral part of the equity and credit decision process. A successful plan conveys the prospects, opportunities, and growth potential of the private company or the nonprofit corporation.

The strategic business planning process for local economic development involves going through a series of steps, the results of which are incorporated in a single document. The plan becomes a "living document" and is updated, usually on an annual basis, as the business grows and expands. The overall thrust of the plan is to outline the strategic intent and vision of the business; document the goals, objectives, strategies, and tactics to accomplish the strategic intent; and outline the resources needed to achieve the goals (see Figure 2.1).

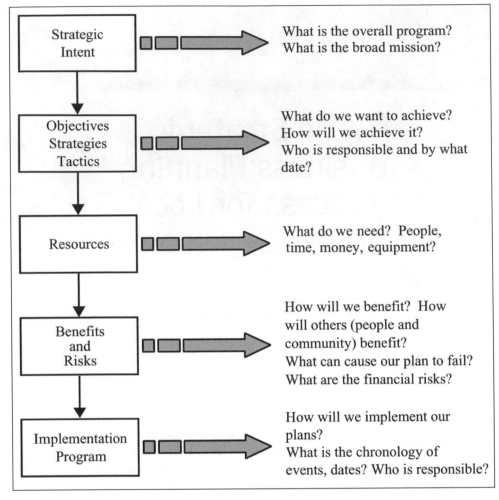

Figure 2.1. Business Planning Process for Local Economic Development

Components of the Plan

The key components of the plan are clearly defined in sections covering all aspects of the company, its key executives, the company strategy, the markets and the competition, a description of the product or service, how the product or service will be sold, the financial information covering the operations of the company and the product, a risk analysis, and an implementation program. The business plan is more than a market and financial feasibility study. It is strategic in nature and explains to the reader how the organization will function to produce the product or service and deliver it to the marketplace.

The outline for the business plan is as follows.

▬ Executive Summary

The executive summary describes the strategic intent or value proposition and all the key elements of the plan in three pages or less. The strategy, market, competition, product or service, management team, projected financial data for the product and organization, outline of benefits and risks, and amount of money requested are all included in the executive summary. Often, the executive summary is the only part of the business plan that is read. Therefore, it should be concise but comprehensive.

▬ The Organization

All organizations have backgrounds and strategies, whether they are partnerships, nonprofit corporations, or community development corporations (CDCs). It is essential that the reader understand what the organization is and where it is going. Therefore, the plan must present a complete description of the company, its key executives, and its strategy for growth and development of a product or service. This section of the plan includes a discussion of how the group or business is organized, the key members of the management team, and the roles and responsibilities of each member. It also includes a discussion of the strategic intent.

The plan should also discuss the shortcomings of the management team and how you will augment the skills that may be lacking internally. Your recognition of the need to pool resources with others will demonstrate to the lender or venture capitalist that you have thought about this issue and are willing to take steps to meet the overall needs of the company. Resumes of the key executives and any outside resources you will use should be provided in the appendix.

▬ The Market and Competition

The section on the market and competition provides documentation and graphics of your findings from the market analysis (see Chapter 4). Lenders and investors will examine this section of the business plan to ascertain if, in fact, there is a sufficient market for your product or service.

Everyone who develops a product believes that the market is strong and that his or her product will sell. But we know that this is not reality. Therefore, the market and competition section analyzes and presents an economic approach. It describes the market for the product, the size of the market, the existing and planned competition, and the overall direction of the market.

The financial institutions, foundations, and grant makers will want to know what will motivate buying decisions in the market and whether you can segment your market and quantify sales by segments. Finally, your investors or lenders will want to know how you will position your product or service and what your defensive strategy will be if the market turns or a recession occurs. It is in this section that you will discuss your sales strategy and explain any variation in pricing that may occur with a change in market conditions. If you plan to reduce or increase prices on the basis of a change in market conditions, these changes should be documented and explained in the financial pro formas.

The section should also discuss how your product will be sold, who will be responsible for the sales program, what selling methods you will use, and how you will promote your product or service. It should quantify the expected sales on a monthly, quarterly, and annual basis.

▬ The Product or Service

The product or service should be described in detail. The challenge in this section is to describe the product or service and analyze it in terms of its features and costs—especially in comparison to the competition. Whether the product is an apartment building, a technology center, or a day care center, the business plan must outline the key features of the product or service and the cost of these features. For example, you have determined that an apartment project is needed in the market area. You have outlined the key features of the units as well as the costs, but you have found that the costs will require that your rents be $75 per month over market rents. If these features are critical to your project, you must convince your lenders, partners, or investors that you can deliver a fully leased building and that the features will not affect absorption of the units.

In this section, outline the development issues that may result in delivery delays and competing projects. Explain whether the regulatory process will affect your project and whether community support will be a positive or negative factor. Also explain how your project or service will be "leading edge" and different from other projects currently in the market.

▬ Financial Statements and Forecasts

The financial section of the business plan is the last section to be prepared, although the risk and implementation sections follow this section in the format. The details in the financial section will vary depending on the organization and funding required. For economic development projects, we recommend that a "start-up expenses" spreadsheet be assembled for 1 year. Revenues and expenses should be forecast for a 5-year period. (See Tables 2.1 and 2.2.) The financial pro forma, or *discounted cash flow analysis,* for the product or service should be forecast for 5 years, as presented in Chapter 5.

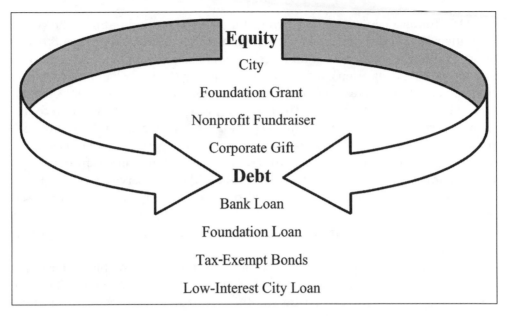

Figure 2.2. Layered Financing for Local Economic Development Nonprofit
Venture

The first step in the financial section is to state a funding request. How much
money is needed and for what purpose must be outlined concisely. Be consistent
in explaining how the money will be used and over what period of time. In determining the requested amount, provide for operational reserves in case costs rise
or additional funds are required to keep the company in business through the
completion of the project. It is important to provide for cash reserves so that
the organization does not have to return to its lenders for additional funds before
the project is completed.

The next step is to describe the equity and debt required for the project. Equity
is the risk money raised by selling ownership interests in the business or the
project (see Chapters 5 and 6). Equity investors understand that there is no required repayment of these funds. They typically seek higher returns than traditional investments. They will see these high returns only if the project is successful. Most equity investors view investments as long term. Therefore, the company
can feel fairly confident that the money will remain available if the company performs according to the terms outlined in the business plan. The forms of equity
capital vary according to the type of investor. Equity can take the form of cash,
mortgage-backed securities, venture capital, venture funds, or stock offerings.

Unlike conventional projects, local economic development projects often
require multiple layers of debt or a series of low-interest and conventional loans,
outright grants, foundation grants that may require repayment, tax-exempt bonds,
and incentives from public agencies. Whereas a conventional project may have
one to three equity sources and one form of debt, the nonprofit or community
project debt and equity structure will have many sources, as shown in Figure 2.2.

Debt funds are raised as borrowings for a specific period of time. They carry with them the burden of payment at regular intervals, with fixed payments over a short or long period of time. Although there is a fixed repayment schedule at a stated interest, debt often carries with it a requirement for collateral or security that will affect the interest required on the loan. For example, in real estate, the collateral is the land and building. The loan is typically referred to as a nonrecourse loan, meaning that the lender will take the property as collateral and not seek repayment from the borrower if the loan goes into default. In other cases, the lender may take shares in a company, require personal guarantees from the borrowers, or request advance payments to be set aside in a reserve account with the lender. Established lines of credit may also be required by the lender to be used as an offset against cash flow. Debt with venture capitalists usually comes with equity features that permit the venture capitalists to share in the upside performance of the company.

Capital requirements for the corporation and the project will be determined once the start-up costs and 5-year operating expenses are estimated. These are added to the equity and debt needed for the project and service to arrive at a total of funds needed to launch the company and implement the program.

Capital requirements of the plan begin with a set of assumptions for cash, investments, loans, and operations. Beginning with a forecast of cash flow, we prepare a detailed schedule of cash inflows and cash outflows for 5 years.

The steps to determine the start-up costs and operating expenses involve verification of costs in the marketplace and an estimation of revenues that will flow to the organization once the project is in place. The costs include locational costs (the costs of occupying space to conduct operations) and operation costs (the costs of initial inventory, equipment, utilities, salaries, taxes, benefits for staff, local taxes and licenses, legal and accounting fees, insurance, etc.). Each organization, depending on its mission and program, will have special expenses typical of its industry.

All costs should be itemized and compared to market costs for your industry. In some cases, you may find it better to retain consultants rather than hire staff. Consultants usually charge more than full-time employees, but their services may be more feasible for your needs initially. Comparison shopping for the facility, personnel, and equipment is necessary for new companies. The emphasis is on keeping the costs down so that the organization can focus on the project or service and complete the program without seeking additional funding.

Revenues will be a function of sales or fees, depending on the program. In the day care center, fees will be derived from enrollment. In a condominium project, revenues will be obtained from sales. For an apartment development, revenues will come from lease payments. In the start-up company, revenues will lag behind start-up costs and initial operating expenses. In estimating cash flow from operations, it is critical to understand that project revenues will not finance the operations initially. Rather, cash flow from investments and grants will support the organization initially (see Table 2.3). These are itemized separately on the cash flow forecast.

Loans and financing activities are also listed separately so that the potential lender or investor may view the company's financial management activities. These loans or activities include borrowings under the line of credit, proceeds from issuing long-term debt, repayments of debt, and payments of dividends.

Tracking the cash flow is a necessary tool for capital management. Itemizing the increases and decreases in cash monthly, quarterly, and annually will enable the organization to manage and plan its activities and production. Any potential shortage of cash can be forecast in advance so that the company can manage effectively.

For the purposes of the economic development program only, we recommend an abbreviated income statement rather than the more detailed accounting approach that itemizes cost of sales. Unless the economic development project is a specific piece of equipment rather than a development program or service, analyzing cost of sales is irrelevant to the overall financial stability of the program. Therefore, the income statement will set forth the sales or fees from services, operating expenses, and income from operations similar to the pro forma set forth in summary form and in greater detail in Chapter 5 (see Table 2.4). The goal is to identify the net operating income for repayment of debt that has been used to start the company. In Table 2.4, we assume that the combination of debt and equity will enable the nonprofit corporation to be self-sustaining over time.

In Table 2.4, both the for-profit and the nonprofit corporations will deliver medical services to the elderly. The first-year revenues are $2 million for each entity. We have assumed that credit losses will reach 10% of total gross revenues. Operating expenses will vary for each entity because the nonprofit will assume greater start-up costs to secure financing and retain appropriate staff. Both groups will lose $75,000 from operations the first year. Interest expense, however, will be higher for the for-profit entity because its funds will come from conventional sources carrying a combined interest rate of 12.5%. Rather than outright grants, the for-profit corporation will also secure foundation loans that must be repaid within 2 years. The nonprofit corporation will save on interest expense because it will layer its financing and the combined interest rates will not exceed 9%. At the end of the first year, the for-profit corporation will have $777,500 to repay the foundation loan and amortize its debt. The nonprofit will have $743,000 to undertake the same financial obligations. The for-profit corporation will use any losses to offset taxes. The nonprofit need not consider the tax obligations. Both corporations must have sufficient cash flow to continue operations without returning to their lenders.

The final table provided in the business plan is the prospective balance sheet. The balance sheet aids the investor in evaluating your ability to manage your assets (see Table 2.5 for an example).

Investors use various financial statistics to evaluate companies. *Return on equity* (net income/total equity) is a profitability measure that investors can compare to that of similar companies to evaluate their potential return on their investment. The *current ratio* (current assets divided by current liabilities) is critical because it measures the company's ability to meet short-term obligations. If the

(text continues on page 27)

Table 2.1 Start-up Costs for Community Business Service Center (in $000s, Before First Day of Operations)

Start-up investment	60
Initial inventory	20
Building acquisition	65
Machinery and equipment	40
Furniture and fixtures	20
Remodeling and decorating	10
Licenses and permits	5
Prepaid insurance	3
Utility deposits	1
Preopening advertising	2
Professional fees	
Legal	2
Accounting	2
Other	—
Total Investment for Operations	**230**

Table 2.2 Projected Operating Expenses (in $000's)

Category	Total Year 1
Salaries and wages	150.0
Employment taxes	25.0
Employee fringe benefits	30.0
Rent/mortgage	10.3
Telephone	1.2
Electricity	1.0
Gas	0.5
Water	0.5
Office supplies	1.0
Postage and delivery	1.0
Maintenance and repair	1.0
Local taxes/licenses	5.0
Legal and accounting	4.0
Insurance	3.0
Vehicle expense	2.0
Advertising and promotion	1.0
Supplies	25.0
Total Operating Expenses	**261.5**

Table 2.3 Variables for Cash Flow Forecasts

Cash Flow From Operations
Cash received from customers
Cash paid to suppliers
Salaries and wages paid
Interest paid and received
Miscellaneous receipts
Net Cash Provided From Operations

Cash Flow From Investments
Capital expenditures
Proceeds from sale of property and equipment
Decrease (increase) in other assets
Net Cash From Investments

Cash Flow From Financing Activities
Net borrowings from line of credit
Proceeds from issuance of long-term debt
Proceeds from issuance of short-term debt
Repayments of long-term debt
Repayments of short-term debt
Repayments of dividends
Net Cash From Financing Activities

Net Increase (Decrease) in Cash

Table 2.4 Comparison of First-Year Income Statements of a For-Profit and a Nonprofit Corporation (in $)

	For-Profit	*Nonprofit*
Gross revenues	2,000,000	2,000,000
Less: credit loss or vacancies	(200,000)	(200,000)
Effective revenues	1,800,000	1,800,000
Operating expenses (Including start-up expenses)	(760,000)	(820,000)
Income (loss) from operations	(75,000)	(75,000)
Interest income and/or expense	(187,500)	(162,000)
Net operating income before loan repayments	777,500	743,000

Table 2.5 Prospective Balance Sheet for the Brava Valley Community Business Service Center (in $000's)

Assets

Current assets		
Cash	175	
Investments	25	
Accounts receivable	60	
Inventories	80	
Total		340
Property, plant, and equipment	125	125
Total Assets		**465**

Liabilities and Stockholders' Equity

Current liabilities		
Short-term debt	25	
Accounts payable	25	
Income taxes payable	20	
Accrued liabilities	20	
Total current liabilities		90
Long-term debt	95	95
Total liabilities		185
Stockholder's equity		
Common stock	60	
Paid-in capital	55	
Retained earnings (deficit)	165	
Total stockholders' equity		280
Total Liabilities and Stockholders' Equity		**465**

Statistics

Return on equity	19.13%	Profitability measure
Current ratio	3.78	Company's ability to cover its current liabilities with current assets
Debt-to-equity ratio	0.66	Creditors are providing $0.66 of financing for each $1.00 provided by shareholders
Net worth	$280	Total assets less total liabilities— the company's real worth after paying off all obligations

current ratio is greater than 1.0 (and hopefully much greater than 2.0), the company is signaling the investors that there are sufficient assets to cover current liabilities. *Debt-to-equity ratio* (total liabilities/total equity) indicates the extent to which the company is leveraged. The higher the leverage, the greater the risk inherent in the investment. Net tangible worth (stockholders' equity less intangible assets such as good will) represents the safety factor for creditors. Creditors and investors alike will measure a company's net tangible worth to evaluate the company's ability to sustain losses. A high net worth indicates the company's ability to sustain losses. Combining the financial reports, creditors or investors can calculate if the total funding request is sufficient for the company to survive a downturn in the economy.

The creditors and investors can also calculate the reasonable length of time the company needs to generate profits and pay off debts.

▬ Benefits and Risks

An often omitted section of the business plan, the benefits and risks are critical for the company to evaluate and to present to creditors and investors. The business plan should outline clearly how the program or service will benefit the investors, community, and local businesses. Outlining if and when the investors will have capital returned is an option. If the company is unsure of the timing of profits and if profits will be distributed or reinvested into the company, the business plan may explain these facts in very general terms.

The business plan should state how the program, project, or service will benefit the community and local businesses. For example, if a 300-unit apartment project is built by BVCC developers, how will the community benefit? How much additional property tax revenue will be generated from the land development? How much will the occupants of the apartment project spend in the community (for retail sales, for services, etc.)? How many new jobs will be created for the community? How will the project positively or negatively affect the transportation services in the downtown? The goal is to quantify the benefits so that the readers—creditors, investors, local community groups, local government—will understand the importance of the project.

At the same time, the business plan should outline the risks associated with the project. For example, will the local government be forced to provide additional fire or police protection? Will the creditors and investors assume risks if there are construction delays? Will construction delays affect the leasing program? Will the company have sufficient funds, or will it be required to seek additional funding 2 or 3 years from now? The risks should be explained in detail. At the same time, the company should outline what steps it will be taking to mitigate the risks.

▬ Implementation

The final step in the business plan is to detail how the company and program will work. Specifically, the business plan should outline the entire implementation program, with target dates for each component. The implementation program is more than a time line. It is a step-by-step approach to putting the "deal" in motion. How will the group with the dream as outlined in Chapter 1 put its organization together? How will the land or building be acquired or renovated? How will the product design begin? What are the first and last steps the company will undertake when it receives its funding?

In essence, the implementation program is not only the conclusion to the business plan but also the first step in making the program or service a reality. The investors and creditors need to understand how the company will operate, what steps will be taken, when progress reports will be received, and when the program or service will be completed.

▬ Summary and Conclusion

The strategic business plan is the preferred vehicle used to secure investment for local economic development. As a rational approach, it is that mental creation needed before the delivery of a program, service, or product. It provides the opportunity for a company to learn about its industry and market. It increases the probability for success by enabling the company to secure a competitive advantage.

Case Exercise

1. Once again, the BVCC board of directors has asked you to help them understand how the strategic planning process will work for them. Building on your needs analysis, prepare a short presentation outlining the broad spectrum of possibilities for this program. From your needs analysis, provide a preliminary list of potential uses for the land area, together with benefits and risks that may exist. This is a brainstorming effort, so push the boundaries to think of all the possibilities.

2. As part of your effort, review Tables 2.1 through 2.3. Take your most likely uses, and build the inputs for the start-up costs, operating expenses, and cash flow forecasts.

Organization for the Venture and Program

The strategic business planning and financial approach forces us to focus on venture selection, community needs, and priorities. Throughout the financial analysis, we have been oriented toward goals and objectives of the community to deliver a sound economic development project. The venture works well within the community, it has a market niche, and it can be financed through a variety of funding sources. Invariably, as the team works through the market and financial analyses, changes in the venture or development project are proposed. It becomes essential for the community to return to its original focus and dream to ensure that the goals are met. Revisiting the needs, priorities, and beneficiaries is a first step. The second step is to develop the legal entity or organization to deliver the project and possibly other projects.

The criteria for project selection will vary depending upon the type of venture being considered. Whether the project is a social project with a local government, a community-owned enterprise, a joint venture with a private entrepreneur, or a large-scale industrial manufacturing complex, scope and resource requirements will be different. In each instance, however, the community decision makers must know what the deal points are, what the community is willing to put into the deal, and what benefits they seek. Several factors should be considered by the leaders of the organization when developing criteria for economic development projects.

Organizational form determines destiny. In many cases, the form of the organization predicts, enhances, limits, or restricts the options for action. For example, a community development bank must act like a bank, no matter how community oriented it is. Banks cannot, by law, own the businesses to which they loan money—at least not under U.S. laws. Selecting an organizational form is usually done long before venture selection, so it is important to understand which form will work best.

The form that a business takes will often depend on how big it is and what it can and will do. A large manufacturing plant is typically not owned by an individual. On the other hand, small corner grocery stores are rarely owned or operated by large corporations. But this is not always true. With the explosion of the Internet, there are many exceptions and many possible new business forms. In some cases today, the business form is a hybrid of several forms combined through an array of legal and financial agreements.

The form a business takes is an important issue for the provision of goods and services because it determines

1. How the business is financed
2. Who owns it
3. Who controls its activities
4. What the main objectives are likely to be

These are important considerations for a community interested in community economic development or the provision of social services. We will examine the range of organizational forms and the rationale for each in the next sections.

Corporations

To incorporate means "to form a body." A corporation is a body or person created by law that acts in the business world as a single entity. That is, a corporation can sue or be sued, borrow money, negotiate contracts, and do many things a person can do.

Because a corporation is not a real person but an entity that is permitted to exist by various state laws, legal permission to start a corporation must be secured. This is usually done by articles of incorporation filed with the proper state agency. Such articles must list the purposes for which the corporation is formed, whether it is profit or nonprofit, its place of doing business, and the names of its officers.

Bylaws for the corporation may then be established that will detail who the owners or members may be, their responsibilities, and any other important matters, such as permanent committees, a paid executive officer, and the responsibilities of the executive director and board of directors. Because both articles of incorporation and bylaws tend to be standard, they can be prepared by a lawyer very easily. Models are also available on the Internet.

A corporation may be formed by one person (in some states), only a few people, or a great number of people. The important aspect of the corporation is that it acts as one person, no matter how many owners or members exist. Each owner or member's liability for debts or losses of the corporation is limited by the legal entity.

A corporation also has other important characteristics that specially define it, such as

1. Special taxing arrangements
2. Vehicles for the accumulation of capital
3. Governance by a board of directors elected by corporate shareholders; boards are solely responsible for setting policy and hiring key corporate employees
4. Continuing existence with an easy succession because ownership is transferable

There are basically two different kinds of corporations, for-profit and nonprofit. Although both are characterized by limited personal liability for executives and by governance by a board of directors, nonprofits have special taxing arrangements set forth by Federal Guidelines under Section 501(c)(3) of the Internal Revenue Code. Either a for-profit or a nonprofit form may be more suitable, depending on the project.

▬ For-Profit Corporations

The primary goal of for-profit corporations is to provide a solid return on the investment for shareholders. Shareholders are owners who have risked their own money to purchase stock in the corporation. A corporation is financed by selling shares to people or organizations interested in owning a part of the company and sharing in its mission, control, and profits. This provides the corporation with start-up capital for operations and/or expansion.

Each year or quarter-year, the corporation may pay dividends from corporation profits to its shareholders to recognize the value of their investment. These dividends are a return, like interest, on money that the investor allows the corporation to use. This is one way for the investor to make money on the shares. The other is to transfer the shares in the open market or via a restricted sale to others.

For-profit corporations also provide a return through capital appreciation or an increase in the worth of the business. For example, a corporation may start with land, buildings, and equipment worth $1,000,000 and sell shares of stock at $10 a share to raise money. If the value of the land, buildings, equipment, and company sales rises and the business as a going concern has steady sales, receivables, and strong leadership, the value of the company may rise from $1,000,000 to, say, $4,000,000. Investors owning company stock are likely to sell shares for considerably more than the original $10 paid.

As share price rises, capital increases. The shares may be sold to other shareholders, in some cases back to the corporation, or to whomever is willing to pay. Many new companies whose chances of success are uncertain, as well as companies that cannot afford or do not choose to pay dividends to investors, use the goal of appreciation as a way to encourage others to invest. The best example of this phenomenon is the formation of new Internet companies that sell shares at high opening prices to eager investors but make no promise to pay any dividend or even to create a salable product.

For-profit corporations have numerous advantages for investors and individuals interested in starting enterprises. First, they provide a means to raise a substantial amount of capital without incurring the need to pay it back immediately or at all. Second, the corporation has enormous tax advantages for the company and most of its investors. Further, the corporation provides its shareholders both upside potential and limited downside risks. Though an investor can make money via the investment, he or she is not personally liable for the debts of the company. The downside exposure for the investor is the initially paid investment amount to purchase shares.

The feature of limited responsibility for shareholders in corporations is an important consideration for people in low-income communities when business ventures for economic development are under consideration. The fear of losing all they have if the business fails may prevent people from involving themselves in a business venture. If a business is started as a single ownership, a partnership, or a loose association of people, losing everything is exactly what can happen. If the business is formed as a corporation, the only money risked is the money spent on the business itself. Of course, the investors can lose the funds that they used to purchase shares in the corporation. However, they cannot lose more than their investment.

Stock ownership in a corporation has yet another advantage. It provides shareholders with the opportunity to control the activities of the corporation as owners. In theory, the control of a for-profit corporation rests with its shareholders or owners. Each owner has as many votes in the corporation as the number of shares that he or she holds. Through these votes, the owners have the right to make all decisions about the corporation's business. They can decide to reinvest profits in the business, return profits to the shareholders through the payment of dividends, invest in community services, or contribute to a political campaign. Through shareholder votes, the shareholders can voice their approval or dissent over the direction of the company, the payment of executives, merger and acquisition options, and other significant items submitted for shareholder approval. However, the day-to-day decision making rests with the corporate officers (Figure 3.1).

▬ Nonprofit Corporations

Like for-profit corporations, nonprofit corporations are characterized by limited personal liability, opportunities for capital accumulation, and governance by a board of directors. Unlike for-profit corporations, however, the purpose of nonprofits is not to offer investors a return on their investment but to advance

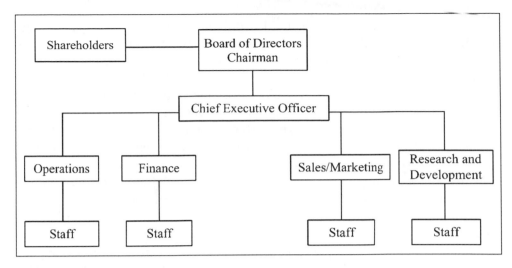

Figure 3.1. Private Corporation Organizational Model

social or community good. Nonprofit corporations are established for educational, charitable, and cultural purposes.

Nonprofit corporations may obtain funds for start-up costs and operation in a variety of ways. Foundations, individuals, government agencies, private agencies, and businesses may provide grants, donations, or contributions. The nonprofit corporation may engage in certain kinds of activities for which it is paid. It may invest its money to earn dividends or interest that can be used to support the activities of the corporation.

In many ways, nonprofits can operate very much like for-profit corporations, and some nonprofit corporations do. Most private universities, for example, are nonprofit corporations. They receive money from donations, tuition, and dividends and interest on investments they make. They spend these funds to provide classrooms, teachers, libraries, and other educational services. However, unlike for-profit corporations, nonprofit corporations cannot use the money they make to provide profits to individuals. This difference is the key to several aspects of the nonprofit corporation. Because it cannot provide profits to individuals, it cannot sell shares of stock to raise money. Consequently, the nonprofit corporation does not have owners in the same sense that a for-profit corporation does. It has members, and who they are is usually broadly defined by its articles of incorporation and established by its board of directors in bylaws.

A nonprofit must be recognized by the Internal Revenue Service under Section 501(c)(3) and the equivalent state bureau of corporation to conduct business within the broad statutes and be exempt from taxes.

Because nonprofit corporations cannot raise money by selling shares of stock, they usually rely on grants, donations, and contributions to get started and to continue their activities as well. The special taxing arrangements that nonprofit corporations enjoy make it possible for them to receive gifts of money.

For-profit corporations typically do not have these arrangements, although the tax code does specify that individuals and organizations that make contributions can deduct the value of these contributions from the income on which they must pay taxes. In some cases, deductions are limited by the tax code. For those individuals and organizations able to reduce the amount of taxes they pay, this deduction is a powerful incentive to make contributions.

An additional tax advantage is available for nonprofit corporations. After filing their articles of incorporation, they can apply for tax-exempt status. Nonprofit corporations do not have to pay taxes on any monies they receive either through contributions or through their own money-making activities, such as providing direct services to their members, as in the case of a school or church.

Special tax arrangements for nonprofit corporations have particular advantages for community groups. Community organizations that conduct their activities through a nonprofit corporation can receive grants, gifts, and donations tax-free. This allows the community organizations to keep all of the monies earned for direct support of their community and business activities.

As in for-profit corporations, individual members and board members of a nonprofit corporation have limited personal liability. For example, if a corporation creates debts, its members are not liable for those debts, and their personal savings or belongings are not jeopardized.

To retain nonprofit tax status, there are some activities in which a nonprofit corporation cannot engage. Nonprofits do not receive direct financial gains from the activities of the corporation except for the wages they receive as employees or for the scholarships or stipends received for educational purposes. More importantly, nonprofits cannot engage in political advocacy. They cannot contribute money to political campaigns, take public stands on political matters, or spend money in an attempt to influence the public or legislators on political issues. In many cases, nonprofit organizations sponsor activities that may seem to be political. For example, voter registration and voter education are exempt from these prohibitions. Though it may appear that the nonprofit League of Women Voters, the National Association for the Advancement of Colored People, or some churches may be exceeding their mandate, their actions can usually be legitimately described as voter education.

▬ Community Development Corporations

In the past three decades, an increasing number of community or neighborhood corporations have been established in communities across the country. About 1,000 of these are groups specifically designed as community development corporations (CDCs). In most instances, these are neighborhood councils and associations that have functioned together for some time and have developed some common goals and strategies that have economic development objectives.

CDCs are specifically designed for economic development. These organizations aim to demonstrate that poor and low-income people can operate their own programs. Communities that have an active membership united by common goals, strong leadership, support, and involvement from the community residents have been able to incorporate themselves like other groups to solve their own problems without personal financial and legal liability. Incorporation provides such groups with a legal vehicle to limit members' responsibility for the activities of their organization. In so doing, it puts them in a stronger position to bargain as nonprofits with funding agencies for grants. Grants provide the necessary program funds, and operational funds came from project operations.

A CDC may be interested in providing jobs and job training to people in the community. But it will also be concerned with having sufficiently skilled employees to run and operate the business successfully from the beginning, which may limit jobs and training for community people. The CDC will be interested in using profits from the business, if any, to reinvest in the business to strengthen and expand it. But it may also be anxious to use the profits for programs and services needed by the community or to return dividends to owners. Its profits will probably not be sufficient to do many of these things.

What the CDC is, who it hires and fires, how much is paid, and what is done with the profits from the business are all decisions that must be addressed by local leadership. Members and the board of a CDC will have to be in agreement among themselves and quite clear about which goals are most important and which ones they are willing to sacrifice in favor of the primary ones. It may be necessary to use the skills and assistance of outside technical experts. In addition, there will be a manager to run the corporation's activities on a daily basis. He or she will need to be free to do the job without constant interruptions and advice from members and the board. But to make decisions that are consistent with the CDC's objectives, he or she will have to understand and agree with these objectives.

As one might anticipate, the board of a CDC is usually structured to guarantee that low-income residents retain a majority of voting seats. In some cases, it is very desirable to provide board seats to representatives of outside groups of professionals, businesspeople, local government officials, labor union leaders, and others. These groups do not control the board. In most instances, the outside members can be restricted in number or given nonvoting membership.

Most CDCs have relied heavily on grants and federal funds for operating monies but also have a strong interest in running social and housing programs for the neighborhoods they represent. CDCs run social programs like day care centers, job training centers, community service centers, and other neighborhood services. More importantly, CDCs provide a mechanism to create new and much needed businesses for low-income communities. These businesses range from grocery stores to auto repair facilities to restaurants to incubator or start-up technology businesses. In addition, they either start or provide support for cooperatives, credit unions, buying clubs, and housing management companies. Figure 3.2 shows a typical organizational structure for a CDC.

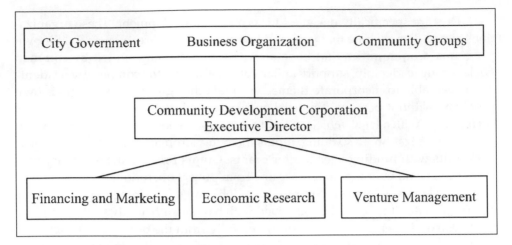

Figure 3.2. CDC Organizational Structure

The potential for the CDC is to create wealth. There is no question that the creation of wealth and jobs is the primary goal, even in instances where secondary goals also exist, such as providing generous employee fringe benefits, offering equal employment and training opportunities, or contributing to worthy social causes. In the final analysis, all of the important decisions in such businesses are always oriented toward trying to make a profit, no matter who benefits.

Cooperatives

When people do things together in cooperation, they can often accomplish more than when they act alone. Cooperatives are perhaps the oldest form of industrial organization in the world. We know from very early biblical records that groups organized before the birth of Christ to sell the product of their labor or pool their income to buy products from others. Today, cooperatives are legal institutions that are business organizations owned, operated, and/or patronized by a group of people working together for their common benefit. The cooperative's purpose is usually to provide members with services and products they need and want at lower prices or to provide members with better prices for things they make and sell. Co-ops usually have the following characteristics:

1. *One person, one vote.* Typically, co-ops get money to start by selling shares to potential members. A share, really, is exactly that. The owner of a share has a share in the ownership of the co-op, a share in the responsibility of running it, and a share in the benefits of the co-op's activities.

2. *Open membership.* This means that anyone can join. Some co-ops, however, may have membership by place of occupation or other criteria. This may be a desirable thing to do in co-ops that are started in low-income areas for the benefit of residents in certain neighborhoods. By limiting membership to people who live in those neighborhoods, the co-op can guarantee their control over its activities and policies.

3. *Small return on investment (on dividends or shares owned).* Return is usually limited to a small amount (less than 5%). Because most co-ops are established to save members money by lowering prices, or to increase members' income by offering higher prices for their products, profits are kept to a minimum. In addition, the members of profitable co-ops often vote to keep the profits instead of distributing these profits to the members. In this way, the co-op pools resources that can be used for expansion or to provide extra services to its members.

4. *Patronage refunds sometimes given.* These are payments of money refunded to members at the end of the year and based on the amount of business a member has done with the co-op that year. Whether or not the membership votes to issue patronage refunds raises the same considerations as profit distributions do. Refunds are helpful to low-income members but are likely to be small.

Consumers' Cooperatives

Consumers' cooperatives are started by groups that need resources that cannot be purchased individually and that want to price them at reasonable and affordable levels. In isolated rural areas, for example, service co-ops have been formed to provide water, electricity, and even telephone services when big companies do not supply them. One of the most common types of co-ops formed is a food store, although this is not usually the most profitable kind of cooperative.

The important characteristics of these co-ops are the following:

1. *Their owners are their customers.* Although most co-ops have nonmember customers, the most important customers are usually the shareholders. That means that they have a hand in deciding what products will be sold and in deciding about other activities designed to meet their dual markets.

2. *Their primary interest is in access to quality goods at low prices.* A consumers' co-op benefits people by giving them their money and giving them a say in what the co-op does. This lack of emphasis on profits may be a disadvantage for groups that are interested in accumulating capital to start other projects. Some capital may be accumulated, but this is not the goal.

▬ Credit Unions

A credit union is a special type of cooperative. Its members save money together and make loans on each other's deposits at low interest rates. Like any financial institutions that handle other people's money, credit unions are regulated by federal and state laws that describe how they must operate, how they can accumulate money, and what loans they can make. And as special types of cooperatives, some of them have regulations much like those that characterize co-ops:

1. *One person, one vote.* Credit unions take in money for their activities and loans through the savings of their members. Usually, any person who deposits savings in a credit union becomes a member-owner with one share for each $5 deposited. Most credit union charters state that regardless of the number of shares one person owns, he or she still has only one vote in determining the policies and activities of the credit union.

2. *Association bonds and open membership.* Credit unions' members must have some common or "association" bonds that unite them. This may mean that they work at the same company, belong to the same club or association, or live in the same neighborhood. Once the union common bond is established, anyone who belongs to that group may join the credit union by depositing savings with it.

3. *Return on investment (or dividends on savings).* At the end of each fiscal year, most credit unions pay their members dividends or interest on the amount of savings the members have with the credit union. Credit unions may simply be an activity for the benefit of their members, or they may be a financing resource as well. So the issue of whether to distribute dividends to credit union members becomes important in low-income communities. Members like to receive the dividends as concrete proof that their savings are earning money. If the credit union retains and pools the interest instead, the indirect benefit is greater to members in new credit unions. The credit union can also be used in this way to accumulate capital in the community for further economic development.

4. *Refunds on interest paid on loans sometimes given.* Many credit unions' members who have paid interest on loans receive interest refunds at the end of each fiscal year. This occurs if the credit union has surplus earnings and the members or board of directors vote to return some of the earnings in refunds on interest payments. In most cases, such refunds are credited to members' accounts. Whether this is done raises some of the same considerations as dividend distribution.

Why start a credit union? Low-income people have very good reasons for starting credit unions, even better reasons than wealthier people. Many low-income people are unable to borrow money from banks and other lending agencies.

Furthermore, with bank consolidation, most low-income areas simply do not have any financial institution other than check-cashing stores.

Traditionally, credit unions have made loans to members for the purchase of consumer products and services. Most credit union charters specify that loans should be made for "provident and productive" services. This has typically meant loans to pay old bills with high interest rates; pay taxes and medical expenses; buy furniture, appliances, and automobiles; and pay for weddings, vacations, and funerals.

However, there are a few instances in low-income communities where successful credit unions with a substantial supply of money to lend have interpreted as "provident and productive" loans for various economic development projects that would increase the supply of money in the community rather than decrease it through consumer loans.

CDC-based credit unions have begun to look for ways to lend money to low-income and welfare families for the purchase of their homes on a cooperative basis rather than lending them money to pay rent to absentee landlords. For example, with four families sharing a multifamily house, the credit union issued loans to each of three members of each family to help purchase a shared home.

Credit unions are formed on the basis of a "common bond," as stated in the credit union charter. The common bonds that unite many potential credit union members may be

1. *Occupational*: Members work for one employer or are in the same trade.
2. *Residential*: Members live in the same community or neighborhood.
3. *Associational*: Members belong to a common association or club, such as a church, a labor union, or another cooperative.

Most experts agree that for a credit union to be effective, there should be a potential membership of at least 100 people.

Credit union charters are issued by various agencies of the state governments or by the Bureau of Federal Credit Unions. A board of directors and officers may be elected, and the credit union may begin collecting savings and making loans to members.

The Single-Owner Business or Sole Proprietorship

The promotion of new small enterprises is increasingly seen as the major vehicle for low-income community recovery. Furthermore, the small single-owner business has always been the most common form of business in this country and is familiar to everyone. Almost every community has a large number of single-owner service and retail businesses such as local grocers, beauty shops, barber shops, shoe stores, cafes, bars, and restaurants. Everyone who is in business for him- or herself is a single owner or single proprietor. This includes farmers, retailers, artists, craftsmen, handymen, software web designers, and small manufacturers. Businesspeople usually choose this form of organization when the proposed business is small, when one person can operate it alone or with the help of the family or a few employees, and when sufficient savings exist or a small loan may be obtained.

The single-owner business has a great deal of appeal to many potential businesspeople for several reasons:

1. *Control of the single-owner business's risks and rewards rests with the owner.* He or she does not share control with partners, shareholders, or boards of directors. If the business is small and has few employees, the owner should be able to stay small and unencumbered. The owner is close enough to the operations of the business to make most of the major decisions. In general, control is limited only by the laws of the community, the requirements of creditors, or the requirement of the major brand name company if the business is a franchise.

2. *In addition to having complete control of his or her business, the single owner also has the sole rights to any profit he or she makes.* After the bills of the business are paid and obligations to lenders are fulfilled, the single owner keeps all the profits and does not have to share these with partners or shareholders.

Because the single owner has control over profits, he or she is solely responsible for the business and debts or losses of income. He or she is responsible personally for the business. That means that if the business has debts that cannot be paid, the owner must pay them by using any profits, savings, or other funds. The owner has to secure his or her own sources of capital, using personal resources as security. The use of personal resources for security is the primary risk to the single owner's security and that of the family. Security is a greater risk than the money invested in the enterprise.

No matter how large or small, single owners have to consider the following issues:

1. The location of the business must be checked to make sure it is zoned for business. Many people who start businesses in their homes find that the use of their home is not legal because the home is not in an area zoned for business. Residential offices or residential businesses may have restrictions on the number of people visiting the premises for business.

2. If the business name is not the name of the owner, it must be registered as "doing business as a fictitious business name" so that business creditors will locate the original owner if the business goes into debt or fails to pay its bills.

3. Licenses and fees are required for business in virtually all businesses, and a "permit to do business" is required. These permits are issued by a city office. Certain kinds of businesses may have to be licensed to operate. This often includes barber shops, beauty shops, restaurants, and liquor stores.

4. Bonding and insurance are necessary in businesses that deal with the potential acts of the company or its employees that risk the public's trust or may cause some form of damage. To be bonded, the company must be able to prove that it has enough assets to complete jobs it has been hired to do. Bonding is needed in many construction businesses. Individuals may need to be bonded, too, if they handle large sums of money. The single owner will need insurance for his or her premises and to protect his or her employees who handle large sums of money against possible loss or theft. The reason for these regulations is to fix the responsibilities of the firm and to protect the public from unsafe and unsavory businesses or illegal business practices.

Why start a single-owner business? As part of an economic development program, the single-owner business has serious limitations. These businesses are usually small and unprofitable. They do not provide significant employment opportunities or opportunities for capital accumulation. They do not provide ownership opportunities except for one person, and even if they are successful, there are seldom many ways in which the total low-income community can benefit from those successes. For the low-income individuals who may consider the option, the difficulties in financing a single-owner business and in having to risk the loss of personal as well as business property may make it a poor choice. But there are times and situations when it may be considered. Some of these are the following:

1. *When specific funds are available for single-owner businesses.* Certain loan funds have been created by businesses and other private sources for profit businesses. Many local development agencies run by cities and associations of businesspeople are making such loans available to privately owned businesses. If such a loan can be found to start a business that provides products or services to the community and it takes people off welfare or retains local income, then single-owner enterprises should be considered. The Small Business Administration, a government agency, provides such loans. It finances single-owner businesses and provides small firms with technical assistance.

2. *When a good franchise opportunity or a sheltered market is available for a single owner.* The franchise may provide the owner with training and experience that will be useful for future business. The single owner with past experience may then be able to go out on his or her own. If he or she has good credit and a history of past business experience, the single owner may be able to put together a profitable business and will then have more control

over the business than if there were partners. The single business owner may eventually incorporate the business when it grows, providing increased ownership, income and employment opportunities for the community, and taxable advantages for the single owner.

Partnerships

When two or more people go into business together, the simplest form of organization they can choose is a general or limited partnership. Partnerships exist in many kinds of businesses. Retail and service establishments, as well as manufacturing and wholesaling businesses, are often owned and operated by partners. In most cases, these partnerships are small. But real estate brokers, stockbrokers, lawyers, and doctors commonly enter into partnerships for their business operations. In these cases, the partnerships are often large and very profitable. At the community level, the partnership is typically used for real estate development.

Community groups such as churches, schools, and CDCs are permitted and even encouraged under federal tax laws to build new housing for low-income groups and seniors. The government extends tax credits for such activities. Nonprofit corporations cannot use the tax exemptions or tax credits. However, if they partner with for-profit corporations, they can pass the tax credits and tax exemptions to the for-profit partners. Often, CDCs will enter into a partnership arrangement with a development company that can bring the construction expertise and also use the tax credits or other tax exemptions to offset taxable income.

For tax purposes, a partnership files an information tax return with the federal and state governments. This states the income of the partnership after expenses and the share of each partner. Then, the partners each pay income tax on what they have received as personal income. The business itself does not pay any income tax. But it may be liable for property, Social Security, and inventory taxes.

Why start a partnership? As part of an economic development program, partnerships, like single-owner businesses, have some limitations. Most are likely to be small and unprofitable, providing few employment opportunities and little chance for significant capital accumulation. Small business partnerships are likely to involve only a few partners and do not provide widespread ownership possibilities. Because partners have unlimited responsibility for the affairs of the business, a partnership may be risky for low-income groups. In addition, problems of personal conflict and disagreement may often make a partnership a difficult form of business operation. But there are advantages to sharing responsibility and liability for debts and losses and in having the skills of more than one person involved. An example of structure and responsibilities for a public/private partnership is shown in Figure 3.3. The private development corporation shares responsibilities with the local government partner.

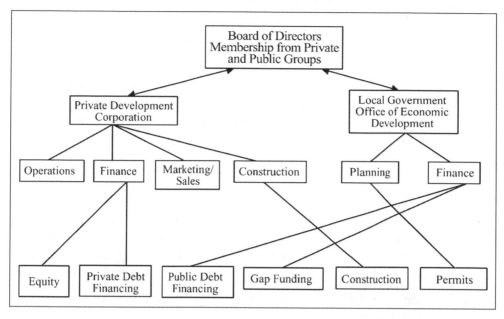

Figure 3.3. Public/Private Partnerships: Organizational Structure and Responsibilities

If a business grows, it may be possible to dissolve the partnership and re-form it into a corporation to limit liability (see the previous section "Corporations"). But the dissolution is a legal proceeding that may be time consuming and may interrupt the business. Depending on the laws of the state and the form of the partnership agreement, dissolution may also be necessary on the death or withdrawal of a partner.

Several reasons for considering a partnership are as follows:

1. *If tax credits and/or loan money* or a franchise is available, a partnership may be able to prove that it has among its members the necessary business skills. The group as a whole is stronger than a single individual.

2. *If an individual needs additional money* or skills from others to enter a business without a loan or outside technical assistance, the partnership structure will facilitate the process, thereby drawing upon the skills of the group.

3. *If the partnership form is common or required* in the particular business, it will be beneficial that the group operate as such to achieve its objectives. In many states, professionals such as doctors or attorneys can form limited-liability corporations. However, they may also operate as partnerships. In some other kinds of businesses, such an arrangement is customary.

4. *If the number of people who want to be involved is relatively small*, the partnership structure works adequately. Partnerships can be made up of many people as well as just two or three.

Conclusion

The organizational form will ultimately dictate the financial and implementation strategies for the project. Consequently, the format must be developed carefully so that options and opportunities for financing and development are not precluded. The overall goal is to maximize the opportunities to obtain the broadest base of financial support available.

Case Exercise

You have conducted your needs analysis and have selected a venture for BVCC. The board and staff have approved your recommendations. On the basis of what you know about BVCC's qualifications, its need for funding, and its desire to accomplish its goals, outline the optimum organizational strategy to implement your program.

1. State the type of organization you will use, and prepare an organizational chart showing how the organization will function.

2. Describe the roles and responsibilities of each officer. If you are using a public/ private partnership model, describe how the partnership will function.

3. Discuss the financial arrangements. For example, will tax benefits be passed from one group to another? How will you approach the financial component of the project?

Finding the Market

In financing local economic development, equity investors or lenders consistently use market analysis as the initial phase for underwriting a project. Understanding what is required rather than what is desired is the key to success in economic development finance. Of course, we want to push the boundaries to include creative and innovative projects that serve as catalysts for larger and more comprehensive local economic development. However, the first step is to identify the market risks—where you are and what is required. Knowing the historical market performance, calculating the present conditions, and forecasting the future will define possible local economic development options. This chapter will focus on market analysis, using an analytical approach to feasibility.

Market Analysis

Market analysis is not an exact science. It is a combination of indicators used as one measure for investment. In evaluating the market for projects or services, supply, demand, price, and location are key components to be measured. These components are measured within the context of a market area and demographic patterns.

The *market* is defined most simply as a business's customers. Individuals, households, other businesses, public agencies, private clubs, and institutions compose a market. Each may purchase products and/or require services. The market is that geographical or program boundary within which goods and services flow to customers. The program boundary for senior citizen services, for example, may be defined one way, and the boundary for the delivery of retail goods may be defined another way. In an advanced information age, geographic limits are much more subtle and difficult to determine. However, for most projects, some geography is usually the base point from which the analysis begins. The market boundary also includes demographic patterns and clusters typically based on population, age, income, and education.

If we think of market analysis as a pyramid with the analysis of demographic patterns and clusters as the base, we can then layer analyses of other variables that will help us reach a conclusion as to the feasibility of our program. For a community supermarket, we might provide a geographic boundary of 10 miles, based on discussions with supermarket chains. Therefore, to our analysis of demographic patterns, we will add a layer of analysis seeking data for spending patterns. For a preschool program, we will add data specifically isolating the number of children under 5 years but will also examine work patterns and journey-to-work data for the family households with children under the age of 5 years.

In our illustration below, we want to understand the retail gap in a community. Using a market-serving index, we calculated the opportunities within several geographic boundaries after we analyzed the demographic patterns.

$$\text{The Market Serving Index} = \frac{100 * \frac{(\text{\# of businesses by retail category in the market})}{(\text{\# of all businesses in the market})}}{\frac{(\text{\# of businesses by retail category in the county})}{(\text{\# of all businesses in the county})}}$$

A market-serving index less than 100 indicates a potentially underserved category (i.e., the share of businesses in the market is less than the share of similar businesses within the county), whereas a market-serving index greater than 100 indicates a potentially overserved category (i.e., the share of businesses in the market is greater than the share of similar businesses within the county). When you are evaluating the results using this type of analysis, opportunities may exist from a supply-side perspective, but other factors, such as demand, social, regulatory, and political factors, may make certain opportunities viable and others not. In our illustration, we assume that the business mix should reflect the business mix of the market area. Where gaps exist, we assume that these are niche opportunities (see illustration and Figures 4.1 and 4.2).

Our illustration allows us to focus on those retail segments that are generally underserved in each market area, as determined on the basis of our market-serving indexes using the formula described above (Figure 4.1). From there, we can isolate specific opportunities in the apparel and accessory retail sector that show the opportunity for establishing a market niche. Therefore, from Figure 4.1, we know that furriers and fur shops, maternity shops, handbags and leather goods, miscellaneous apparel and accessory stores, custom tailors, bridal shops, family clothing stores, women's clothing and boutiques, and western apparel shops could possibly be supported by the community. Depending on the demographic profile of the community, these retail areas may be needed or may be lacking because of income and age distribution of the population.

To validate a specific market niche, we must create additional layers by focusing on the population mix within each of the geographic units and by examining

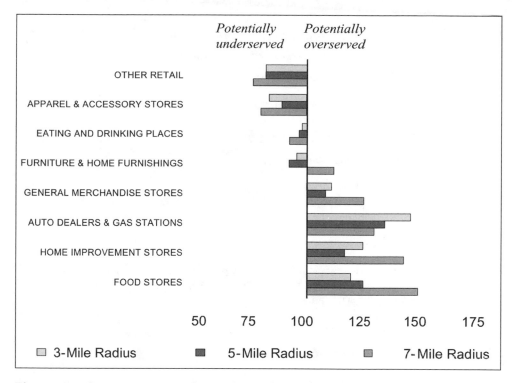

Figure 4.1. Opportunities in General Retail

competition and the proximity of our selected local economic development area to that competition.

The primary and secondary market areas will vary according to our local economic development program. The primary market area for a custom tailor may be 1 to 3 miles, but the primary market area or "trade area" for a super regional shopping center may be 1 hour's distance. The primary market area for a downtown office building could be all or part of the central business district, depending on highway and transit lines.

Market areas can be defined by compiling primary research through face-to-face interviews with merchants, community groups, civic leaders, lenders, and real estate professionals. Physical surveillance of an area is essential to determine living, shopping, employment, and transportation patterns. Market areas can also be defined through secondary research or materials that are provided in reports or publications of others. These reports are frequently available through public agencies, banks, trade associations, and government agencies (Wurtzebach & Miles, 1994).

Demographic patterns are identified so that goods and services can be targeted to "customer groups." These groups are defined by population, age, income, and employment. Each has distinguishing characteristics, profiles, and needs that

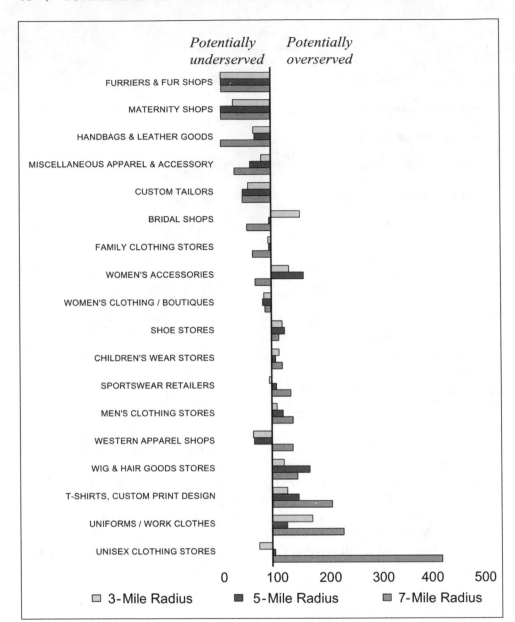

Figure 4.2. Opportunities in Apparel and Accessories

vary from one place or region to another. For example, the single-head household will require different housing, shopping, employment locations, and recreation patterns than the dual-income family household with children. Understanding the profiles of each group is necessary to measure the overall demand for the planned project or service and enhance the marketing strategy. "When developing a marketing strategy, you will have to decide what market segments you will use to define the scope of your market and [ensure] that they all work together

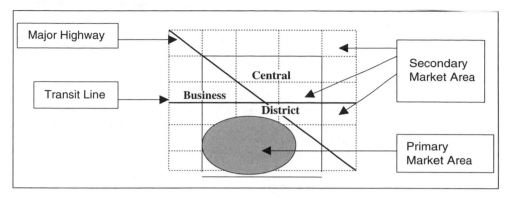

Figure 4.3. Primary and Secondary Market Areas, With Census Tract Grids and Major Arterials

toward a common goal of determining the total potential market of your product or service" (Wurtzebach & Miles, 1994, p. 50).

Data to determine demographic patterns are located in U.S. census materials in libraries or on-line. The U.S. Census Bureau provides population and housing statistics, employment profiles, and age and income characteristics by census tracts, cities, counties, and states. Demographic statistics are provided by regional planning agencies and state agencies. Typically, the analyst will use the lowest level of data, such as census tracts, and then aggregate to the geographic level of the market area unless the primary or secondary market area is a larger geographical unit, such as a city or county.

Figure 4.3 illustrates how a transit line and major highway affect the size and shape of the primary market area for prestige office space or upscale apartments. Although *central business district* has been defined and incorporated in local government documents, the major highway and transit line, in effect, reduce the size of the primary market area for premium office space or luxury apartments to less than 50% of the total area. The transit line and highway serve as a boundary locking in a defined geographical area because their configurations virtually cut neighborhoods and communities and isolate these markets from nearby communities. The secondary market area, however, includes the entire grid or broader area where limited competition and prospective tenants may be located.

Through access to U.S. census data, local or regional planning agency demographic and employment reports, and Bureau of Labor Statistics reports on county business patterns or consumer surveys, we can build a profile of the population in the primary and secondary market areas. Data from these documents will provide statistics on the local population—age breakdown, number of households and average household size, ethnic origin, income, and employment status—as well as information on the trade area, including current gross sales and per capita sales by store category (Table 4.1).

The analysis tools used for evaluating the U.S. census data are likely to be presented in block groups, tracts, or aggregate numbers. The data measurements can be accessed on the city, place, or state levels. The U.S. census data can be accessed

Table 4.1 Primary and Secondary Data Sources

	Primary	*Secondary*
Neighborhood	Churches, temples Interviews with brokers, tenants, merchants, shoppers School districts Police, fire departments Neighborhood associations	Neighborhood papers U.S. census Planning agencies National data sources Military
City	Building permits Planning agencies Business licenses Industry councils	Telephone directory Chambers of commerce City business directories Service organizations Military Ethnic Yellow Pages Newspapers
State and Regional	Legislative documents State board of equalization Public utilities	U.S. census County business patterns Regional planning State employment development

through the Internet at www.census.org. The data presented in Table 4.2 outlines demographic data in our Brava Valley community based on the 1990 U.S. census and estimates for 1994 and 2001. U.S. census data will provide further breakdown of age, income, and employment. Using these breakdowns can enable an organization to specifically target its product or service to certain age groups or income groups.

Market Dynamics

Understanding the market dynamics relative to population and employment is essential. As we continue to add on the layers of analysis of our pyramid, we must analyze the relationships between and among the population groups, how employment is or is not driving the economy, and what potential niches can be supported by the demographics of a community. For example, Brava Valley employment is expected to increase 9.3% from 1994 to 2001. Though the largest increase will come from agriculture, there are strong employment gains in the managerial, administrative, and professional and technical sectors. These gains, although small, indicate a change in the employment profile of Brava Valley and an opportunity for Brava Valley to seek businesses that can be supported by these workers (Table 4.3).

Table 4.2 Brava Valley Population/Employment Summary, 1990-2001

1-Mile Trade Area	1990	1994	2001 (Est.)
Population	220,000	280,000	338,000
Households	95,000	183,000	195,000
Average household size	3.5	2.7	2.8
Race			
White	180,000	200,000	225,000
Black	20,000	25,000	27,000
Asian	15,000	31,000	52,000
Hispanic origin	4,000	15,000	22,000
Other	1,000	9,000	12,000
Median age	32 years	27 years	23 years
Income			
Per capita	$17,000	$25,000	$33,000
Median household	$30,000	$36,000	$46,000
Average household	$28,500	$34,000	$44,000
Employed population	97,500	97,400	106,000

Table 4.3 Brava Valley Employment Projections, 1994-2001

| | Annual Averages | | Absolute | |
Occupation	1994	2001	Change	% Change
Managerial and administrative occupations	6,090	6,930	840	13.8
Professional, paraprofessional, technical	21,220	24,180	2,960	13.9
Sales and related occupations	12,430	13,940	1,510	12.1
Clerical, administrative support	15,760	17,170	1,410	8.9
Service occupations	17,050	19,280	2,230	13.1
Agricultural, forestry, fishing	860	990	130	15.1
Production, construction, operations, materials handling	23,770	23,780	10	0.0
Total, all occupations	97,180	106,270	9,090	9.4

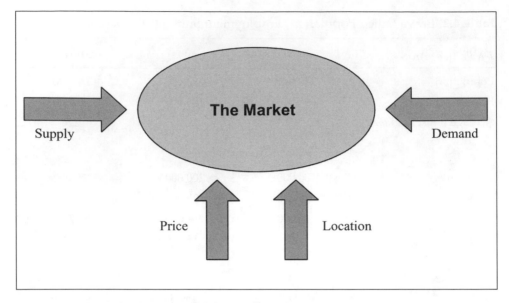

Figure 4.4. Key Components of the Market

In addition to the employment changes, Brava Valley's population is also changing. Though total population from 1994 to 2001 is expected to increase 20.7%, the largest increase will come from the Asian population (67.7%), followed by those of Hispanic origin (46.7%) and others (33.3%).

As we continue to analyze the economy further, we can view each of the economic sectors in terms of growth. By comparing the locality with its region and each of the economic sectors, as illustrated earlier in this chapter, the relevant sectors that will match the growth in population and employment can be identified.

The Key Components of Analysis

The critical element in the financial success of a local economic development project is identification of the potential customers. The key components—supply, demand, price, and location—drive local economic development and, ultimately, all the financial factors that influence investment. If these key components are analyzed and the appropriate market niche is identified and used for the local economic development project, financial success will occur. Therefore, one must have a complete understanding of the key components and the market structure (Figure 4.4). Understanding of the structure includes accurate, detailed knowledge on (a) the competition, (b) the comparative advantage of the project or service over the competition, (c) the length of time to absorb the demand, and (d) the market factors creating a "push-pull" effect.

━ Supply and Demand

Using the market area data—geographical boundaries and socioeconomic statistics—we are ready to analyze the market in terms of overall supply and demand. Supply is created by investment and measured by the number and type of similar projects or services in the market area—both existing and planned-for development. Supply is further refined in terms of competition, price, and location. Price and location will differentiate a project or service from its competition.

Demand is measured by need—what users want and what they are willing to pay. Wurtzebach and Miles (1994) defined effective demand as follows: "Potential users of proposed space not only must exist, but they must also have the purchasing power to acquire (through purchase or lease) the desired space-over-time plus associated services" (pp. 20-21). The process of finding a need to fill can be narrowed by concentrating on the needs of the population, corporations, and public agencies with whom the organization comes into regular contact. The most successful business ideas are the ones that fill newly recognized needs for an existing client base.

Because all economic markets fluctuate over time, demand will increase or decrease. The level of demand can be measured in several ways. The first measure uses *vacancy rates* as a proxy for demand and can also be used as a proxy for too much supply. Vacancy rates can be measured by the number of square feet or vacant units in competitive projects within the defined primary market area. For example, if vacancy rates are greater than 5% to 8%, demand may be low because prices are too high or because the supply in the market is high.

The second measure is *absorption rate*—that rate at which new or existing units or square feet of offices for certain businesses are leased or sold in a defined period of time in the primary market area. In tight markets, 100% of all new business space or housing units may be absorbed. In more fluid markets, 85% to 95% of new space may be absorbed, thereby leaving a hangover of space for the following defined time period.

A third measure of demand is *household or employment need*—that specific percentage of the target population who will use a medical office facility, request counseling from a mental health provider, purchase coffee from a new coffeehouse, or visit and purchase books from a new bookstore. This measure of demand relies heavily on local market surveys and focus groups to identify what households and businesses need. If the survey is conducted through statistically random sampling techniques, results can be applied to the total market population. If the survey is conducted ad hoc without statistical sampling, the results can only describe what some of the target population is seeking. The latter method is often used to describe the benefits or risks of a local economic development. Valid statistical sampling is the preferred approach for ensuring that the target market and market niche is highly accurate. However, valid statistical sampling is expensive and requires professional survey researchers to undertake the process. Therefore, the simplified approach is more often used with local economic development projects.

Illustration: Sample Market Testing Survey

Using either a statistically random approach or the simplified approach, the market testing survey will enable the professional to determine the more precise demand and need for a given product and service. The market survey will also enable the professional to focus on a pricing structure. For example, a potential business offering jitney-bus services to and from a shopping center might develop a list of survey questions and follow through by conducting surveys in a residential neighborhood, downtown areas, or local business districts. Surveys can provide detailed information on "wants and needs" that are often overlooked through quantitative analysis using population and employment statistics only. A questionnaire could include the following:

1. Do you live in this neighborhood?

2. Do you work in this neighborhood?

3. Do you ever shop at ABC Shopping Center?

4. How often do you shop there in a week?

5. Do you shop during the week only?

6. Do you shop on the weekends only?

7. Do you shop during the week and on the weekends?

8. If a jitney-bus service were available to transport you to the shopping center, would you use it?

9. How frequently would you visit the shopping center if transportation were available?

10. Would you visit the center more often than you do now if a jitney service were available?

11. How much would you be willing to pay each way for this service? $.05 $.10 $.20 $.30 $.50 $1.00

12. How often do you think you would use the service at each of these prices?

13. What other places besides ABC Shopping Center should be served by the jitney?

14. What would you be willing to pay to go to these places?

15. How often would you go to each?

16. Do you think other members of your family would use such a service? If so, who and how frequently?

17. How would you define "really good transportation service" for this neighborhood? Please be specific.

— Price and Location

Price and location are the two variables that affect the viability of projects or services. As defined above, price is the value that the consumer places on a good or service. A monthly rent of $1,200 for a one-bedroom apartment will be afford-able for one group of consumers and unaffordable for but desired by a larger con-sumer population. The $300 one-bedroom apartment is likely to be affordable for a larger consumer population but undesirable to many because of its location. A $25-per-hour charge for the use of a computer in a cybercafe will attract fewer cus-tomers than a $5-per-hour charge. Understanding the price/location component in the supply/demand equation is essential to building the foundation for the financial analysis and the business plan for local economic development.

— Capture Rate/Market Share

In balancing out the supply/demand/price/location equation, the *capture rate/market share* becomes the final measure used to evaluate the market condi-tions. The capture rate or market share is the percentage of the competitive market that your project or service is expected to obtain. In products or services, a 3% to 5% capture rate of market share is the minimum required and accepted by lenders (this figure is based on discussions with the lenders over the past 8 years). Inves-tors in a technology product or medical service will expect at least this level. How-ever, in a local economic development project, lenders and investors consider a 3% to 5% level ideal unless your project is the only project of its kind and no other similar projects are planned for the primary and secondary market area. With the length of time required to construct a project, many variables may change, thereby affecting the market share forecasted by the developer. The lower the capture rate, the less market risk the lender or investor would have to assume.

Venture Selection

When considering possible ventures, it is important that the community organiza-tion share a sense of what projects are important and what resources should be invested, under preferred conditions and investment returns. The shared under-standing should be documented in the form of investment criteria as a reminder to all involved in project planning.

Naturally, the criteria for project selection will vary depending upon the type of venture being considered. A community social project with local government, a community-owned enterprise, a joint venture with a private entrepreneur, and involvement in large-scale industrial or manufacturing enterprises are very differ-ent in scope and resource requirements. In each instance, however, the community decision makers should know what they are willing to put into the deal and what they want to get out of it.

Several factors should be considered by the leadership of the organization when developing criteria for investment in community social or economic development projects:

1. *Project beneficiaries.* Where will the benefits accrue? For example, if the target group is the homeless, then the homeless portion of the program should be clearly larger and useful to that population. A community may rationalize that producing more market-rate housing in the neighborhood will create a "trickle-down effect" so that low-income and homeless groups will have access to the available "old" supply while the upwardly mobile households purchase or rent the market-rate housing. But this concept does not work in reality. If providing housing for the homeless, the community must target housing for that population group.

2. *Projects.* What projects will serve the community? All community activities, whether projects or business ventures, should be based on the same goals, objectives, and framework. Local economic development projects should be selected for their consistency with the development goals of the community and the potential benefits that return to the target community, whether in terms of jobs, skills training, or the provision of goods and other services. The local community organization must share a vision of the important activities and the resources that will be needed to develop the activities. The activities and resources should be documented in a formal list of "investment criteria" that can be reviewed once the final business plan is prepared.

When the project beneficiaries and the projects are identified and the key market components are analyzed, the decision makers will be in a position to evaluate, refine, and document the development program.

Marketing Concept

Marketing concept is a popular term used in business to describe a particular way of delivering services or goods to the market. The marketing concept dictates an examination of the market to determine the niche for the product or service. Once the product or service is determined and selected, the techniques to deliver the concept to the marketplace must be developed. For example, when a retail specialist decides to open a cybercafe, the concept of the cybercafe must be disseminated to the local residential community and to the business community. Issues such as price, location, rent, inventory, and customers have been decided. The marketing concept will translate the vision or goal of the storeowner/operator and the local economic development community to the community at large. The new business will provide access to computer equipment on an hourly basis. Families, individuals, and businesspeople will all have access to the cafe.

The decision to open this new market area for shoppers requires not just an interest in offering it but a marketing plan on how to communicate the business concept to the general public so that revenues and profit can grow while the community is served with access to computers in a safe environment.

Irrespective of what the new business venture is, the investment and lending community will want to know and understand how the product or service will be marketed. Who will be responsible for developing the marketing plan? Who will provide the marketing materials to the public, and what will these materials consist of? Who will be responsible for sales? How will the sales and marketing personnel be compensated? How will the product or service be delivered to the market? It is important for any new venture to work closely with local economic development professionals and local business groups to "network" and seek support for the business venture. The most successful business ventures have obtained support for the products and services long before either is delivered to the market.

The Development Program

Once the market components have been addressed, an analysis of the product or service must be conducted to ensure that the supply/demand/price characteristics are closely matched with the product or service. The attributes of the site or the program/service are described, and the development program is outlined. Lenders and/or investors will often ask the following questions:

1. *Competition.* Where is it? How will it compete with our project?
2. *Net absorption.* How has the market performed in the latest recession, and how is it performing today? How will downturns in the economy affect the project or service?
3. *Marketing.* How will the product or service be marketed?
4. *Preleasing or presales for construction projects.* How much will occur before the infusion of equity or takeout of a loan? Preleasing or presales vary according to overall economic conditions. In a strong economy, lower equity and preleasing or presales are required. In a weak economy, 50% equity may be required and 65% to 75% preleasing or presales requested.
5. *Service programs.* Where will the program be housed? How will the program be staffed? What is the expected demand for services? Who will pay, and how much will be paid for the services?

Table 4.4 Critical Elements of Market Analysis

Element	Focus
Independent	The market study is conducted by a third party—an independent objective entity.
Conservative	The author follows the steps for market analysis, undertaking a site and area analysis; economic and demographic analysis; competitive analysis of supply; analysis of demand; and a description of development program, including size of project, expected absorption, timing, and pricing.
Due diligence	The author outlines a series of scenarios examining the best case, the worst case, and the most likely case. The author defines the parameters of each scenario, identifying the risks, how the project may be affected, and what steps can be taken to reduce risk or turn the project around.

The development program should incorporate all assumptions on a year-by-year basis for 5 years. Included in the program is a description of the project or service, the size and location, the target market of users, the pricing of all components, the expected absorption or units sold/leased or square feet leased, and the demand for services offered.

Critical Elements

Several critical elements are used by lenders and investors during the first stage of underwriting an investment or loan. Table 4.4 shows the details that financiers will look for in the market analysis.

In addition, lenders and investors will examine the development program and the set of assumptions used on a year-by-year basis. The examination will yield a quick overview of the preliminary net operating income (revenues minus expenses) before any assumptions regarding debt and equity. Before any formal discussions but using preliminary data from the developer/prospective borrower or equity partner, lenders or investors will often compare the net operating income numbers relative to the size of the building or service with other familiar projects in the market area. The comparison is made to educate the lender or investor with similar measures of performance in the market area and is often made unknowingly by the borrower or equity partner.

Outlined below are two case studies. The first is a continuation of the Brava Valley Community Corporation problem, and the second is the Community Apartments case, which focuses on market development alternatives.

Following from the initial Brava Valley Community Corporation case study, your assignment is to prepare a market analysis, as discussed in Case Study 2, below.

Case Study 2: Brava Valley Makes a Decision

After evaluating the work of the BVCC students, the board of directors of Brava Valley Community Corporation has selected four new enterprises. BVCC has chosen home- and clinic-based health care along with a valley-wide, on-demand transportation service, a senior housing program using the air base housing, and a motor repair shop and small business incubator in the air base hangars.

A preliminary needs assessment and community analysis by the college students indicates that a large unmet need exists. For example, at least four of Brava Valley's seven communities have no home health providers. The providers that exist do not reach out into the rural areas. The board has decided to use the old JC Penney's store site as the location for the clinic. However, the agency has very limited experience with health care other than operating its senior citizen housing programs. There is general concern over how this project should be operated and who should undertake the operations and management.

On the other hand, BVCC's large fleet of vans and cars is underused. Director Henry Mohamid-Gonzales has suggested that the fleet be used for transporting people to medical and other services to meet the transportation crisis in the area that BVCC serves. Mohamid-Gonzales thinks the transportation area is ideal for an employee ownership spin-off.

In addition, a home health agency will build on BVCC's strong base of social services for seniors and on the agency's other health-related services, such as a family planning clinic.

The clinic will also provide jobs and career advancement opportunities for local Native American youth. This effort, combined with the transportation program, can prevent Medicare and Medicaid dollars from leaking out of the community. Although the poor are entitled to these payments, they often are not able to use them. Because of a lack of local service providers, the poor must go outside of the area not only for medical services but for many other services too. As a result, the opportunities for jobs and increased income in the area are diminished because when people go out for medical and other services, they spend dollars on retailing and the like outside the community.

As a community-based organization, BVCC cannot take on the military base conversion by itself. Therefore, it is proposing that the agency operate the housing program for the base and convert it to senior housing. BVCC has considerable experience in housing, but it cannot use the more than $10 million in tax credits that the rehabilitation of these units will generate for the corporation. The board is in a quandary over what to do about this matter. The college students have proposed that BVCC take over two of the abandoned hangars at the base. One of them could be used as a motor vehicle repair facility for the BVCC fleet as well as contracting out repair services for the Brava Valley School District, the U.S. Marshal, and the state police vehicles. This strategy would retain within the community the employ-ment, experience, and profits generated by these auto and truck repair services. Just how this deal should be structured to limit the board's liabilities and not infringe on local auto companies' turf is still an open question.

Finally, BVCC has the option to open a new small business incubator in the other hangar and nearby three warehouses. These facilities contain more than 200,000 square feet of space. This space has to be rehabilitated for the new projects, and the board wants a quick feasibility analysis of the potentials of breaking even if they rehab the site and make it suitable for office use. Mohamid-Gonzales is very com-mitted to this project because he can train local youth in rehab with a government grant. According to his calculations, the rehab will be only about $25 per square foot because of the federal wage subsidy in the training program. The board finance committee still wants some sound rent and return numbers as well as an organization plan before taking on this project.

All of these projects are worthwhile because they multiply the jobs and income benefits from the initial Medicare, housing, and military base conversion dollars. With the help of the Brava Valley Community Hospital, BVCC is investigating estab-lishing a medical equipment subsidiary so that it can capture these health dollars and use them to revitalize the downtown health clinic.

Case Exercise

Now that the board has made these decisions, it is important to see how they fit into the priorities to ensure that

1. The BVCC is appropriately insulated from liabilities
2. The ventures attract capital and users
3. The organizational capacity of BVCC is enhanced

Your job is to develop a complete market study of the selected enterprises to ascer-tain the feasibility for BVCC's new ventures described above. You must present your findings in three days. Use market data from any county in the state of New Mexico assigned to you by your instructor for your analysis. Your task is to complete the following:

1. Identify the market demand for the projects.
2. Chart the location of the closest competition, if any.
3. Develop a pricing structure for all of the ventures.
4. Develop a program organization chart along with your new activities. Outline the personnel that will be needed to support all development.

Table 4.5 Development Program Assumptions and Preliminary Financial
Review

Variables	Year 1	Year 2	Year 3	Year 4	Year 5
Number of Sq. Ft.					
Rentable Sq. Ft.					
Vacancy Rate					
Net Leased					
Annual Lease Rate/Sq. Ft.					
Revenues (Net Leased × Lease Rate)					
Annual Expenses/Sq. Ft.					
Expenses (Net Leased × Expense/Sq. Ft.)					
Net Operating Income (NOI)					

5. Prepare your recommendations for the board. Identify the market benefits and
risks of your development program.

   Using Table 4.5, prepare a simplified preliminary financial review for each component before evaluating the benefits and risks of your development program.

Case Study 3: BVCC Community Apartments

Community Apartments is a 592-family housing project that was planned for development at the BVCC site. The BVCC board has decided that it wants to target
housing to lower-income groups. Therefore, the BVCC board has told the Community Apartments developers that they will help them get approvals for their apartment project in another county where housing prices are high and more affordable apartments are needed. They have recommended that the developers go to
First City, where the board members have political influence. First City is located
along a major interstate highway approximately 25 miles from the state capital,
5 miles from major industrial and retail development, and 5 miles from research
parks and a new major medical center.

The developer can purchase 47 acres in the northeast quadrant of First City. The land has been annexed into the city, and the general plan states that the property is to be developed as industrial/commercial. New apartments do not exist in this middle-income community of primarily single-family detached homes. For that reason, the developer and the investors are seeking a zoning change so that the 592 apartments can be developed. The rental units are expected to be priced between $750 and $1,250 per month.

You have been retained by the investor group and the BVCC board. They have asked you to evaluate whether they should use their political and financial influence to assist in securing a zoning change. You have been provided with a three-page executive summary and a series of charts on the following pages.

I. Executive Summary

This report focuses on two key elements: (a) the market for apartments in First City and (b) the feasibility of apartments at Community Apartments. The executive summary, shown in Exhibit 1, details the pertinent data related to key market indexes of housing, employment, and income that affect the future growth and diversity of First City and its economic base, as well as a summary of Community Apartments and the benefits and costs of the zoning change.

The major findings are as follows:

A. Diversified Housing Stock Provides the Labor Pool for Job Growth.

- The movement of businesses closer to the workforce is a common phenomenon that occurs as urban development takes place. Over the years, location analysis has shown that jobs follow housing. In the region, we can review the trends over the past 20 years and see that major economic growth has occurred as a direct result of the expanded housing stock. For example, between 1980 and 1990, the expansion of the single-family and multifamily housing stock in Region East resulted in the growth of the economic base through the growth of new businesses and the movement of major corporations to these cities. The growth and/or relocation of businesses closer to the labor pool can be tracked throughout the state and the entire United States.

- Besides providing quality of life for its citizens, the development of housing occurs to provide a labor pool for businesses. Without a suitable labor pool, the economy cannot grow.

- Employment, households, and income are critical indexes in evaluating the market for multifamily housing in First City and at the Community Apartments site. These indexes must be in balance if the economy is to grow.

B. Community Apartments Will Help the City Meet the General Plan Goals.

- Community Apartments will support the city's goal to expand the economic base by providing a rental housing stock and a household base for both commercial and industrial development.

- Community Apartments will assist First City by providing a diversified housing stock to meet three of the city's General Plan goals:
  - To promote equal housing opportunities and provide a decent home and satisfying environment for all First City residents
  - To promote balanced residential growth to provide adequate services and facilities to support such growth
  - To promote adequate and affordable housing in the city by location, type, price, and tenure

C. Freeway County Job Growth Will Exceed Household Growth, Increasing Pressure on Housing Stock.

- Freeway County is expected to have the highest rate of population growth between 2000 and 2020—an increase of 46%. During the same period, the county's employment is expected to grow 74%.
- Although the jobs/households were in balance in 2000 in Freeway County, jobs have now surpassed households. The gap is expected to widen through 2020, placing additional pressure on the cities and the county to provide a diversified housing base.
- The labor market is projected to increase 14.28% between 2000 and 2005, whereas the growth in households during the same time frame is forecast to be 11.14%. From 2005 to 2010, jobs are expected to increase 17.65%, whereas households are anticipated to increase only 12.44%.

D. Freeway County Requires a Minimum of 1,000 to 1,250 Multifamily Housing Units Annually. First City and Second City Need 222 to 276 Units Annually to Meet Local Demand and Affordability Needs.

- Based on our analysis, the minimum demand for rental housing in First City and Second City between 2000 and 2005 is 222 units annually; between 2005 and 2010, it is 276 units annually if the housing market is to remain fluid and meet the growing needs of the workforce. Freeway County will require a minimum of 1,007 units annually from 2000 to 2005 and 1,249 annually from 2005 to 2010.
- To afford the average-priced single-family house of $150,000 in First City, or other locations in Northern Freeway County, annual household income must exceed a minimum of $40,000 to $50,000, yet more than 33% of First City households and 30% of Second City households earn less than that amount. These households can easily qualify for rental housing payments of $775 per month without the burden of principal, interest, property taxes, insurance, assessments, and maintenance totaling $1,300 to $1,400 per month.

Community Apartments will provide 592 multifamily housing units over a 3-year period in a market area that requires more than 3,000 multifamily units during the same period. The development will provide high-quality rental units to meet the growing labor force.

EXHIBIT 1
Executive Summary

Development Plan
47.22 acres
592 multifamily units
1-2 bedrooms, 1-2 baths
In-unit laundry
Recreation center
Parking

Primary Target Market
Freeway County
Dual-career and single households
Seniors

Target Income Groups
< $50,000 (annual)

Growth in Labor Market
2000-2004   9.3%
2005-2010   17.65%
2010-2020   15.68%

Growth in Households
2000-2005   11.14%
2005-2010   12.44%
2010-2020   10.47%

Annual Demand for Apartments
2000-2005   222 First City and Second City
                    1,007 Freeway County
2005-2010   276 First City and Second City
                    1,249 Freeway County

Minimum Income Required to Rent
$23,250-$25,000

Planned Supply for Apartments
0 units, First City
264 units, Second City

Minimum Income Required to Own
$40,000-$50,000+

Eligible Households for Apartments
First City   1,246
Second City   7,855
Total Freeway County   68,581

City Benefits
Community Apartments fulfills General
  Plan requirements for diversified
  housing stock
Community Apartments provides a labor
  pool for employment growth
Community Apartments residents will
  support planned commercial
  development

City Costs
Staff, council, and Planning Commission time to review zoning
change

EXHIBIT 2

Freeway County Employment Projections, 2000-2004

Occupation	Annual Averages 2000	2004	Absolute Change	% Change
Managerial, administrative	6,090	6,930	840	13.8
Professional, paraprofessional, technical	21,220	24,180	2,960	13.9
Sales and related occupations	12,430	13,940	1,510	12.1
Clerical, administrative support	15,760	17,170	1,410	8.9
Service	17,050	19,280	2,230	13.1
Agricultural, forestry, fishing	860	990	130	15.1
Production, construction, operations, materials handling	23,770	23,780	10	0.0
Total, all occupations	97,180	106,270	9,090	9.4

EXHIBIT 3

First City Competitive Apartment Survey

Project	Number of Units	Unit Type	Monthly Rent	Vacancy Rate	Amenities	Comments
Apartment A	36	2 bd/1 ba	$525	0%	Carports; laundry room	Tenants pay electricity
Apartment B	32	2 bd/2 ba	$675	0%	1,000 square foot units; carports only; washer/dryer in units	Building is 10 years old
Apartment C	104	1 bd/1 ba 2 bd/1.5 ba	$510 $595	1%	Laundry room; covered parking; swimming pool	One unit is vacant and will lease quickly
Apartment D	95	1 bd/1 ba 2 bd/1 ba 3 bd/2 ba	$575 $650 $775	3%	Covered parking; laundry room; swimming pool; hot tub; sauna; resident clubhouse	Built in 1980

SOURCE: Apartment manager interviews.

EXHIBIT 4

First City Existing and Planned Residential Development

Project/Location	Number of Acres/Lots	Single-Family/ Multifamily Rental	Project Status
Existing Projects			
Glen Homes	25/86	Single family	2 lots available
Manor Homes	10/30	Single family	24 custom homes constructed; 6 lots available
Subdivision Homes	63/248	Single family	112 homes completed in Unit I; 71 lots completed in Unit II; 16 homes under construction and 49 lots remaining
Fair Housing	22/110	Single family	97 single-family homes completed; 13 lots under construction
First City Homes	32/126	Single family	10 lots remain available
LaPaz Homes	11.78/106	Single family	Cluster affordable single-family homes; 71 homes under construction; balance awaiting approvals
Bird Development	108/526	Single family	Units 1, 2, 3, and 6 under construction; Units 4a, 4b, and 5 complete
Estate Homes	35/211	Single family	130 units completed, 35 under construction; balance of lots available

Race Homes	22/74	Single family	Rough grading completed between N. First St. and I-80
West Homes	8/22 and 3 duplex	Single family	11 units completed or under construction; balance available

Approved Specific Plans

Park Development	206/809	Single family and multifamily	Maximum 131 high-density units
Southwest Development	141/800	Single family	Planned units are for single family; multifamily housing designated but not planned at this time
Hill Properties	20 acres	Single family	
Tree Development	36 acres	Single family	
School Homes	69 acres	Single family	

SOURCE: First City Project and Development Summary; developer interviews.

EXHIBIT 5

First City and Second City
1997 Estimated Households by Income

Income Category	% of Households	
	First City	Second City
$150,000+	5.59	2.02
$100,000-$149,999	3.92	6.07
$75,000-$99,999	9.70	10.26
$50,000-$74,999	25.82	30.90
$35,000-$49,999	21.97	21.32
$25,000-$34,999	13.58	10.56
$15,000-$24,999	10.89	9.29
$5,000-$14,999	7.10	8.38
Under $5,000	1.43	1.18
1997 average household income	$64,020	$57,417
1997 median household income	$46,602	$49,476
1997 per capita income	$22,620	$21,073

SOURCE: National Decision Systems.

EXHIBIT 6

Multifamily Housing Demand Analysis, 1995–2005

Household Projections[a]

| | Number of Households | | | |
	1990	1995	2000	2005
First City	3,149	3,402	3,781	4,251
Second City	21,779	23,528	26,149	29,402
Balance of Freeway County	88,124	95,200	105,810	118,967
Total County	**113,052**	**122,130**	**135,740**	**152,620**

Growth Rates[a]

Time Period	Rate
1990-1995	8.03%
1995-2000	11.14%
2000-2005	12.44%

Proportion of Owners to Renters[b]

Time Period	% Owner	% Renter
1990-1995	63	37
1995-2000	63	37
2000-2005	63	37

Growth in Households[c]

	1995-2000	2000-2005
First City/Second City	3,000	3,723
Balance of Freeway County	10,610	13,157
Total County	**13,610**	**16,880**

Growth in Renters[d]

First City/Second City only	1,110	1,378
Total County	**5,036**	**6,246**

Annual Demand for Rental Housing[e]

First City/Second City only	222	276
Total County	**1,007**	**1,249**

First City Gateway Capture Rate for 3-Year Build-Out[f]

First City/Second City only	88.9 %	71.6 %
Total County	**19.59%**	**15.80%**

SOURCES: Regional Projections 1996; U.S. Census of Population and Households, 1990 and Updates; National Decisions Systems, 1998.
a. Regional and U.S. census projections.
b. Assumes conservative projections of 33% renter housing.
c. Based on projections from 1990 to 2005.
d. Calculated based on growth in households times 37% renters.
e. Annual demand assumes growth in renters divided by 5 years.
f. Based on county or First City/Second City growth of households only. Assumes 592 units divided by total growth in renters over 5 years. If one 264-unit building is built in Second City, limited competition is expected because the Second City project will be 100% senior housing.

Case Exercise

Your investor group and members of BVCC board are well acquainted with First City's Planning Commission and City Council. The investor group and BVCC board believe that they can lobby the commission and council effectively to secure the zoning changes. However, they are not convinced that apartment development is the correct approach. Should they develop housing or focus their strategy on the development of an industrial park/technology center? What are your general recommendations?

1. How will you address the capture rate under the housing demand analysis?

2. The 131 planned high-density units at Park Development are not scheduled to come on line for at least 10 years. If the city pressures Park Development to build sooner, how will the project affect Interstate Apartments' development program? How will you address this with the investor group?

3. What are the three most critical facts the investor group should provide to the commission and the council if they proceed?

4. If the planning director wants to negotiate a deal with you so that Community Apartments can be developed along with industrial and retail/commercial uses in the same quadrant, what will you advise the investor group?

5. You have just located industrial development data for the market area. The following information was obtained:

   - Growth in the labor market: See Exhibit 1 of the executive summary
   - Total occupied industrial space = 1,200,000 square feet
   - Total occupied and vacant industrial space = 1,416,000 square feet
   - Current vacancy rate = 18%
   - Annual absorption rate = 70,000 square feet per year
   - Total acreage planned for development over 5 years = 3,000 acres
   - Average lease rate in primary market area = $1.40 per square foot per year
   - Average lease rate in primary and secondary market area = $2.10 per square foot per year
   - Primary tenants in market area = technology/hardware; technology/software; biotechnology; medical services; transportation

6. An industrial project will not require a change in zoning or a conditional use permit. How much time do you anticipate saving with this type of use? Can you translate the time savings into dollars saved?

7. In evaluating both the housing and industrial scenarios, prepare a position paper for the strategy you choose, using the data from the executive summary and Checklists 4.A, 4.B, 4.C, and 4.D.

Task: As a proxy, identify a building for sale in or near your community. This building can be used as an industrial facility, an office building, a shopping center, or an apartment project. Complete the checklists, using available market data sources as presented earlier in the chapter. Prepare a summary of your findings for presentation to the board and the investor group.

✔ **CHECKLIST 4.A** Socioeconomic Data

Sources of Data: City/county planning departments, economic development departments, real estate brokers, real estate data services (i.e., urban decisions systems), census data via the Internet, state and federal departments of transportation.

Economic Base of Market Area:_____

_____

5-Year Growth Patterns, Total and Industry Sectors, in %:

1990-2000	2001	2002	2003	2004	2005

Population:

1990-2000	2001	2002	2003	2004	2005

Income:

1990-2000	2001	2002	2003	2004	2005

Employment:

1990-2000	*2001*	*2002*	*2003*	*2004*	*2005*

Wage Levels:

1990-2000	*2001*	*2002*	*2003*	*2004*	*2005*

Transportation Patterns:

Freeway:_____ Rail: _____ Air: _____

Adequacy of Public Infrastructure:

Water:_____ Sewer:_____

✔ **CHECKLIST 4.B** Property Review

Property Name and Location: _____

_____

Property Type: _____ Size: _____ Age: _____

Asking Price: _____ Price/Sq. Ft.: _____ Occupancy Rate: _____

Building and Site Characteristics:

Construction: _____

Design Type: _____

Loading: _____

Building Depth: _____

Clear Height: _____

Ceiling Height: _____

Parking Ratio: _____

Size of Tenant Spaces: _____

Tenants:

Name: _____ Credit Info. _____

Name: _____ Credit Info. _____

Name: _____ Credit Info. _____

Name: _____ Credit Info. _____

Name: _____ Credit Info. _____

Office Finish: _____ Lease Expirations 2000-2005 (Sq. Ft. and %):

Site Coverage: _____ 2000 _____ % _____

Amenities:

Sprinkler System: _____ 2001 _____ % _____

Column Spacing: _____

Dock Height: _____ 2002 _____ % _____

Truck Maneuvering Areas: _____

Grade Level: _____ 2003 _____ % _____

Lighting: _____

Ventilation System: _____ 2004 _____ % _____

Floor Thickness: _____

Upgrade/Remodel: _____ 2005 _____ % _____

Environmental Issues: _____

_____

Other Issues: _____

_____

Comments: _____

_____

_____

✔ **CHECKLIST 4.C** Physical Inspection

General Characteristics:

General Overall Condition: _____

Tenant Directory: _____

Landscaping: _____

Signage: _____

Parking Lot: _____

Truck Turnaround Areas: _____

Common Areas: _____

Building Characteristics and Conditions:

Roof Type/Condition: _____

Roof—Visible Leaks or Staining: _____

Hallways—Flooring, Walls, Ceilings: _____

Number of Offices, Flooring, Walls, Ceilings: _____

Visible Deferred Maintenance: Interior _____

Exterior _____

✔ **CHECKLIST 4.C** Continued

Equipment and Security:

Mechanical Equipment and Capacity: _____

Electrical Equipment and Capacity: _____

Ventilation System and Capacity: _____

Energy Conservation (Yes or No; if Yes, type): _____

Fire Safety Equipment: _____

Security Equipment: _____

Other: "As Builts" Available: _____ Certificates of Occupancy or

Equivalent: _____

✔ CHECKLIST 4.D Market Overview

Market Area Data:

Total Sq. Ft. Built: _____

Total Sq. Ft. Planned for Development (Next 2 Years): _____

Historical Absorption Rate (Number of Sq. Ft./Year): _____

5-Year Absorption Rate (Last 5 Years—Sq. Ft./Year): _____

5-Year Absorption Rate—Existing Buildings: _____

5-Year Absorption Rate—New Buildings: _____

Competitive Properties Sold Within Last 24 Months:

Location	Price	Size	Price/Sq. Ft.	Land Area	Amenities/Comments	Key Tenants	Lease Rate/Sq. Ft. Office Area

Competitive Properties Leased Within Last 24 Months:

Location	Key Tenants	Size of Space	Lease Terms	Amenities/ Comments	Land Area	Rent/Sq. Ft.	NNN or Gross/ Expense Stops

Fundamentals
of Economic
Development Finance

To understand how finance fits into the total local economic development scheme for projects or services, we must first examine the overall system that drives the process. The system includes how we take ownership of our project or service, how we fund the program, and how we tap into the capital markets to secure the required funds. Knowing how the system works will provide us with a road map to implement our program.

Principles for the Players

Ownership, value, financing, and development are key variables in the overall system of economic development.

— Ownership

The basic model for ownership includes possession, control, enjoyment, and disposition. For development projects, *possession* includes the right to occupy and enjoy, together with the right to keep others out. For service projects, possession includes the legal right to deliver specific services. Licenses may be required for the delivery of services. *Control* of development projects deals with the right to alter the property physically and to conduct business according to local ordinances. Control in the delivery of services means the right to legally provide services within specific boundaries and zoning ordinances. *Enjoyment* allows the current owner to be protected from past owners or others who might try to interfere with the living or business rights of occupancy. Enjoyment also means the right to deliver services free from others who may compete with the delivery of similar services. Disposition permits the owner to convey all or part of the inherent rights of ownership to others. The *disposition* of services means that the service delivery agency or professional can terminate the specific services at any time unless the services are bound by a legal contract for a stated period of time. For example, the family counseling center could terminate its services if it ran out of money or if its board of directors decided that the center was no longer needed. For a specific project such as ownership of a single-family house, you could occupy the home, renovate it as long as it met building codes established by the community, keep people out of your home or allow them to enter, and sell the home at any time and for any price the market would pay.

Because ownership comes with this bundle of rights, the owner has the ability to undertake certain actions depending on the legal structure. In local economic development, as in products or services, the form of ownership has certain rights and responsibilities. Simple ownership, community property ownership, joint venture partnership, and other legal entities carry with them advantages and disadvantages for investors. Liability, tax consequences, risks, and returns must be considered along with the type of ownership of the business enterprise. For a more complete discussion of property ownership, readers should consult the American Appraisal Institute's handbook, *The Appraisal of Real Estate* (1992) and Wurtzebach and Miles's discussion in Chapter 4 of their book *Modern Real Estate* (1994).

Value

Value is typically defined as the present worth of future cash flows. Value in local economic development projects may be much more than cash flows. It may also include the increase in quality of life for the local community, the value of the project as a catalyst for future projects or programs, and the recognition that the community can foster the creation of new businesses for its citizens. Value has the power to command other goods or services in the marketplace and is actually the amount of money paid for a good, service, land, or building. In examining and calculating value, three approaches are considered: (a) *market value* (the price paid in the market for comparable projects, goods, or services); (b) *replacement cost* (the actual cost of replacing the project, goods, or services); and (c) *income capitalization* (the present value of future cash flows). The three prices are typically close to one another and are reconciled by the appraiser. Whether the local economic development project is a building or a business enterprise, the income capitalization approach usually carries the greatest weight in measuring value.

▬ Financing

Financing is the vehicle used to enable owners to transfer projects, goods, or services from one owner to another. It is the means to leverage the purchase. It is the vehicle used to close the transaction when there is insufficient cash. In most cases, owners do not want to put much hard cash into an investment and instead secure money in the form of debt or grants.

Financing can take the form of equity, debt, or grants. Each source of funding carries with it specific underwriting and return criteria. *Equity* incorporates the dollars required to launch a project. It is that risk portion of the entire package that increases or decreases with value. Equity is typically provided by the developer, the investors, and/or community groups or interested parties. *Debt* is a short- or long-term loan that carries with it a promise to pay. Debt is provided by financial institutions such as commercial banks, insurance companies, investment banks, and/or pension funds. *Grants or subsidies* are what we define as "free" or "soft" money used to reduce the overall cost of the project. Grants or subsidies may or may not be repaid but carry with them an economic or social responsibility for the project. They are provided by local, state, or federal governments, foundations, nonprofit organizations, and other interested parties such as pension funds or commercial banks in search of community projects. If the grants or subsidies are provided by the private sector, they create the "push-pull" effect in the overall development process, as outlined below. We will elaborate on the specific sources of funding later in this chapter and in Chapter 6.

▬ Development

The development process occurs when the marketplace requires a product, service, or specific development. In local economic development, the public sector says, "We want this project because it will create jobs, revitalize the neighborhood, or serve as a catalyst for future development. It is an important project that will enable our community to generate revenues. We will raise the quality of life for our citizens." The private sector says, "We want to develop the project, but the land and permit process are too costly. We cannot develop unless we receive incentives to write down the overall costs of construction and financing. We will not build the project you need unless you provide the land at a reduced cost and provide low-interest loans, subsidies, or other incentives."

Money and values drive the entire process. In the urban environment, economic activity occurs because of inducements provided by the public sector. In other words, the development process for a project or service will be the "push-pull" between those who want the project (government, community, and neighborhood) and those who can deliver (developers, investors, and financial institutions). Although the market analysis discussed in Chapter 4 defines the size and type of project or service, the value perceptions of the public sector and private sector create the struggle or the push-pull effect.

Value in the public sector is defined in terms of quality of life and increase in the revenue base so that development will create a revenue stream back to the community and its citizens. Value in the private sector is defined as the revenue stream, profit, appreciation, and interest payments that are returned to the developers, investors, and financial institutions. Value for foundations and organizations providing grants is typically defined as social and economic good flowing to the community, its businesses, and its citizens.

The essence of financing economic development, therefore, is the act of combining values and perceptions—reaching for the common ground where the public sector, private sector, foundations, and nonprofit organizations can achieve their collective goals. All parties to the transaction understand the agendas. They choose to bring their total resources together to create wealth, whether for the community, its citizens, the developers, investors, or financial institutions.

Money Sources

Our capital markets in the United States function as the vehicle through which all goods and services are financed. The principal players in the capital markets are the investors, the savers, and various financial institutions, corporations, and governments who ensure that money flows between and among investors and savers. These principal players allocate funds according to return and risk. For example, savers may allocate funds to commercial banks, credit unions, life insurance companies, and investment banks for further investments in securities. Banks, credit unions, life insurance companies, and investment bankers will use the money from savers to make appropriate investments that generate suitable returns for the savers as well as for themselves. Investors such as individuals, businesses, or the government will generally use funds as equity participants in projects so that they may achieve a desired rate of return on their funds.

The principal players are the entities that ensure that money flows between and among all participants. The major tool used by the principal players is investment vehicles. These vehicles take the form of savings accounts, life insurance policies, securitized instruments such as real estate investment trusts and other mortgage-backed securities, home and commercial mortgages, and start-up companies, among others.

Securitization is a field commanding attention today. Securitization is the creation of a transferable instrument from a bundle of assets. These assets may be a large pool of mortgages, business receivables, or an equity pool from a portfolio of buildings. The assets have been determined to be "good risks" by financial analysts. They are expected to perform well over a stated period of time; therefore, they can be priced and sold in the financial markets. Investment bankers organize the bundle of assets for sale to investors who want the security of regular payments without the anxiety that accompanies direct investment in a specific industry or sector.

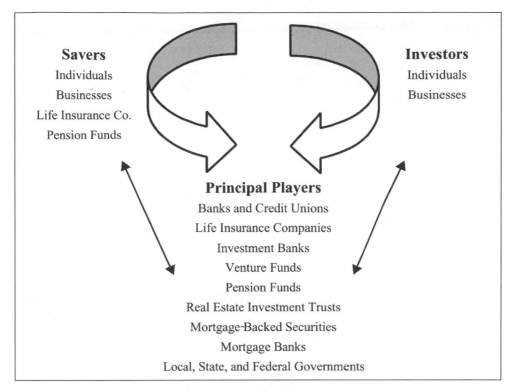

Figure 5.1. U.S. Capital Markets: Flow of Funds
SOURCE: Adapted from Wurtzebach and Miles (1994, p. 376).

When the saver or investor places money with one of the principal players, he or she receives an expected return on the investment. The return may be fixed or speculative, depending on the investment vehicle. In return, the principal player can charge a fee, sell the investment or loan, or use the funds to make other investments, thereby generating returns not only for the saver or investor but also for the principal player.

The capital market principal players are the financing engine for economic and business development. They bring together the funds from the savers and investors and lend money or make additional equity investments. Although they all have their own underwriting criteria or value agendas, they often come together to provide multiple layers of financing for economic development projects (see Figure 5.1).

▬ Private and Conventional Funds

In economic development projects and services, funding will be secured from a variety of sources—private and conventional, public or government funds, grants, or low-interest loans. Private and conventional funds may be generated

Table 5.1 Private Funding Sources

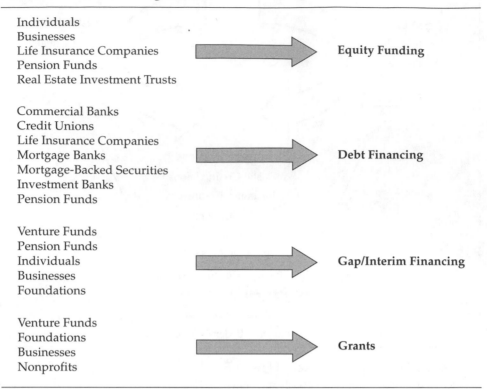

Individuals
Businesses
Life Insurance Companies → Equity Funding
Pension Funds
Real Estate Investment Trusts

Commercial Banks
Credit Unions
Life Insurance Companies
Mortgage Banks → Debt Financing
Mortgage-Backed Securities
Investment Banks
Pension Funds

Venture Funds
Pension Funds
Individuals → Gap/Interim Financing
Businesses
Foundations

Venture Funds
Foundations
Businesses → Grants
Nonprofits

through many sources. The primary equity will come from the developer and/or investors. Equity may also come from pension funds, real estate investment trusts (REITs), life insurance companies, and/or the public sector. At the same time, the investors, pension funds, and life insurance companies may provide permanent financing for a project through direct loans or through REITs and commercial mortgage-backed securities (CMBSs). The public sector—federal agencies, state agencies, or local agencies—may also provide direct loans, although these loans usually carry lower interest rates than conventional loans. Commercial banks, investors, public agencies, grants, and foundations may be sources for interim or gap funding (see Table 5.1).

The underwriting criteria for private and conventional funding vary by institution and type of loan. However, there are common underwriting themes for both the project-specific programs submitted by the developer and the internal portfolio requirements handed down to loan officers through executive policy and a financial institution's credit committee.

Under project-specific and portfolio guidelines, the investor, loan officer, and credit committees will measure expected risk and return through a variety of measures, as outlined in Table 5.2.

Table 5.2 Project and Portfolio Lending Guidelines

Project-Specific Guidelines	*Measures*
Developer or business executives' track record	Historical account of developer projects or executives' experiences
Equity in project	Percentage of equity to debt
Project feasibility	Feasibility study, business plan, market performance
Security of project	Absorption forecasts and dollars in reserve to cover negative cash flow
Security of developer/investor	3 years of certified corporate financial statements; 3 years of personal financial statements
Cost of money	Interest rate, points, letters of credit

Portfolio Guidelines	*Measures*
Ability of institution to lend	FDIC regulations; board policy; credit committee
Composition of portfolio	Existing loans by property type on the books of the financial institution
Location of project	Geographic location
Size of loan and terms	Board policy; credit committee
Evaluation of risk	Subjective with final determination by the credit committee

▬ Public Sector and Nonconventional Funds

Through a variety of financing mechanisms, federal, state, and local governments provide funding for economic development projects and services. State and local funds vary depending on legislative actions by individual states. However, most state governments, through tax-exempt vehicles handed down from the federal mandates, redevelopment laws, and infrastructure financing legislation, enable developers to tap into funds so that projects can be developed.

Depending on the property type or service, government agencies use their powers to provide low-interest loans and gap financing for development. Although the 1990s saw a cutback in funding sources that were typically available in the 1980s, a series of public sources often provide the much needed capital at reduced interest rates to ensure that projects and services are delivered to

Table 5.3 Public Sector Funding Sources

Federal Sources

Tax-exempt bonds
    Low-interest permanent financing for single-family housing, senior housing, apartments, senior medical facilities

Tax-exempt bonds for redevelopment
    Tax increment financing for acquisition, demolition, remodeling, rental assistance, lease assistance

Community Development Block Grant Funds (HUD)

U.S. Small Business Administration loans
    7(a) Loan Guaranty; LowDoc; FASTRAK, 7(M); MicroLoan Program, SBA 504 Loan

State and Local Sources

State mortgage funds for low-interest loans

Community business loans for start-up companies

Technology funds and grants for start-up companies and technology parks

State enterprise zones and community development block grant funds

Tax-exempt bonds for infrastructure development
    Assessment districts, Mello-Roos (California)

Tax-exempt bonds for industrial development
    Low-interest loans for industrial parks, industrial expansions

Revolving loan funds for local economic development corporations

Venture funds for local economic development corporations

Job training funds under Job Training Partnership Act (JTPA), granted by Private Industry Councils

---

communities. Public and nonconventional funds available to economic development projects and services are listed in Table 5.3.

Building the Financial Plan

The integral part of building the financial plan and testing development scenarios begins with understanding the sources and uses of funds, determining the amounts that may be allocated to segments of the project, creating a structure that integrates and allocates funds according to lender requirements, and testing the scenarios. There is no rule of thumb that can be used to split the sources and uses of funds on a percentage basis. Typically, lenders will require between 25% and 35% of equity in a project. The equity requirement is based in part on the lending relationship with the developer or business executive and the sources and uses of other funds for the project.

Monies secured from private and conventional sources together with funds from public and nonconventional sources will provide the overall financial security for the development or service. The building blocks for project finance are (a) the pro forma, (b) the business plan, and (c) the loan proposal. The sources and uses of funds for project finance are presented in Table 5.4.

The *pro forma* sets forth all the assumptions used for revenues and costs. It details how the money will be allocated to cover the cost. The allocation can be adjusted by altering the sources of funds or reducing the costs. The *business plan* sets forth the key factors surrounding the project and program, the organization and key personnel responsible for the project, the market analysis, financial forecasts, analysis of risks, and organizational implementation strategy. The *loan proposal* details specific loan data, sources and application of funds, the collateral summary, the purpose of the loan, the monthly amortization schedule, the business plan, the financial plan, and corporation and individual financial statements. Many aspects of the loan proposal are incorporated into the business plan. Each lender will require its own loan application. The financial plan will be incorporated in the business plan (see Chapters 2 and 7). The building blocks are completed using a conservative, technical approach to ensure that all data are organized and presented to equity and debt participants.

Estimating Capital Requirements

Before the project is started, there must be an estimate of the initial costs and potential revenues. A portion of the initial amount of money for a project is devoted to what are called "start-up" costs. Start-up costs involve many of the activities discussed in Chapter 4, such as determining the market for the project, the size of the demand, and marketing of the project. In some cases, such as new technologies, the start-up costs include the perfection of the product via more research, development, and testing before delivery to the market.

Beyond the start-up phase, most businesses or services require a "break-in" period where advertising and sales have not met the expectation of the organization. All products and services, regardless of the demand, require time for the customer to change to a new product or service. Even in the case of child care, mothers will keep their children with relatives until the new day care center is established and accepted by the community. Therefore, the new enterprise requires start-up capital during the initial period of operations.

Table 5.4 Sources and Uses of Funds for Economic Development Projects

Sources	Uses				
	Predevelopment, Acquisition, Equity, Start-Up	Construction/ Rehabilitation R&D	Management/ Operations	Interim and Long-Term Loans	Subsidies
Equity Investors					
Individual investors	•				
Pension funds	•				
Insurance companies	•				
REITs	•				
Venture funds	•				
Developers	•				
Executives	•				
Conventional Debt					
Commercial banks	•	•		•	
Insurance companies		•		•	
Pension funds	•			•	
Other Private Debt					
Mortgage-backed securities	•	•		•	
REITs	•	•		•	
Investor loans	•	•			

88

Federal Government

	1	2	3	4	5
Tax-exempt bonds		•	•	•	
Tax credits		•	•	•	
Community Development Block Grant funds		•	•	•	•
Small Business Administration	•	•	•		•

State Government

	1	2	3	4	5
State mortgage funds		•	•	•	
Community business loans			•		
Technology funds and grants	•		•	•	
State enterprise zones			•	•	
Tax-exempt bonds	•	•	•	•	
Tax allocation bonds	•	•		•	•

City or County Government

	1	2	3	4	5
Community business loans		•	•		
Revolving loan funds	•	•		•	•
Venture funds	•		•		
Job Training Partnership	•				
Land write-downs	•			•	•
Housing Authority	•		•		•

Foundations and Grants

	1	2	3	4	5
Foundations and Grants	•	•	•		•

Estimating the amount of money and time required is one of the most important decisions for a new venture. Undercapitalization is the major reason that business and social programs fail. Here are the steps that are recommended in estimating capital requirements.

1. *Estimate sales.* Sales are made directly or indirectly. For example, e-commerce businesses offer a service or product. The services or products can be sold directly or indirectly through other Web-based companies. Child care is sold to parents via fees or government transfer payments. It is important to estimate how many units will be sold, to whom, and for what fees. The market analysis will help establish the fee structure. No market is guaranteed indefinitely; therefore, pricing is critical for the new venture. Once you have established the price structure and estimated the market share, you can develop the financial statement documenting the total start-up expenses.

2. *Estimate expenses.* A realistic estimate of expenses along with contingencies is essential. Developing scenarios based on low, moderate, and high costs will provide sufficient variation to cover contingencies. Among the typical cost items are rent, equipment, utilities, supplies, transportation, legal and accounting, permits and licenses, payroll, and payroll taxes. Every business must be prepared to use some of its own capital for expansion and research and development. Failure to undertake the appropriate expansion could result in the delay of "time to market" for product launches or delivery of product. Therefore, the new owner must develop an adequate financial plan to ensure that the required capital is available to meet timely expansions and future contingencies.

3. *Estimate working capital.* Working capital is the money needed to keep the business operating. Early on, the firm or project has to have enough money on hand to pay bills and operate for 6 months to a year without any customers. This amount of working capital can help the new venture deal with such things as slow installation of equipment, a sudden storm, or a slow start-up. It is not possible to know what can happen, so the best thing to do is have enough money on hand to weather any potential storms.

In Table 5.5, we have summarized the start-up expenses and capital equipment requirements for the Brava Valley Community Child and Elder Care Center, to be located in the old JC Penney Building.

Pro Forma Construction

The pro forma is your story. Like a good story, it has to read well and sell well. The investors and lenders will not read the "glossy" brochure. They will immediately

Table 5.5 Sample Start-Up Pro Forma (in $)

	Best Case	Most Likely Case	Worst Case
Capital Equipment			
Computers, printers, fax	10,000	12,500	15,000
Telephone	3,000	5,000	6,000
Play equipment	10,000	12,500	15,000
Furniture and fixtures	85,000	100,000	115,000
Elder care equipment	100,000	125,000	135,000
Landscaping	10,000	15,000	20,000
Indoor plants	1,000	1,200	1,500
Total Capital Equipment	219,000	271,200	307,500
Estimated Revenues	79,815	72,559	65,303
Estimated Operating Expenses			
Payroll and benefits	23,515	21,377	22,300
Payroll taxes	1,860	1,925	2,025
Workers' compensation ins.	475	525	550
Building insurance	780	900	975
Rent	8,760	9,500	10,500
Telephone	455	500	525
Office supplies	5,500	6,000	7,000
Vending supplies	900	975	1,025
Utilities	2,950	3,000	3,050
Legal	1,500	1,500	1,500
Accounting	600	600	600
Cleaning services	1,275	1,275	1,275
Security services	1,500	1,600	1,775
Equipment repairs	1,500	1,650	1,875
Miscellaneous and reserves	3,000	3,500	4,000
Total Operating Expenses	54,570	54,827	58,975
Net Operating Income	25,245	17,732	6,328

turn to the pro forma to examine the assumptions, compare the assumptions to projects or business enterprises with which they are familiar, and test the assumptions against reality. The investors and lenders are looking not only at what the pro forma says and what the investment yields are, but also at what the pro forma

Table 5.6 Pro Forma Structure for Economic Development

1. Assumptions

   Document all **revenue** assumptions, including rents, parking, vending machines, public telephones, laundry, services, revenues from products.

   For development projects only, calculate billbacks from common area maintenance and add to revenues.

   Estimate **expenses.** For development projects, estimate property management, administrative and overhead, common area services and utilities, maintenance and security, real estate taxes, insurance, other expenses, reserves for repairs. For services, estimate rent, payroll, benefits, general and administrative costs, utilities, maintenance, security, insurance.

   Develop assumptions for purchase price and/or start-up costs, equity in the project or business, loan amount, term of loan, interest rate of loan, depreciation, income tax rate, capital gains tax rate, and sales commission or other related business costs. In a service delivery program, it is unlikely that one would determine capital gains tax rate unless the business was sold at a determined date.

2. Test

   Vary and test assumptions with information from development and small business experts, accountants, and attorneys. The marketplace provides strong data to test against the assumptions used in building the pro forma. An automatic fixed escalation in rents and expenses will not be sufficient. Vary the rents and expenses according to historical real data.

   Run the model through before-tax cash flow to determine if variables are sound. If more equity is needed or if the loan amount and/or the purchase price must be reduced, explore methods for securing grants or low-interest loans or for negotiating a reduced purchase price. Often, public agencies can assist with land write-downs if the project is one that the agency advocates.

3. Revise

   Revise the model assumptions based on testing and adding additional data to the equation. Revise for reasonableness, not for a specific return.

4. Final Pro Forma

   After variables have been tested, refine all assumptions and prepare the pro forma, the amortization schedule, the project plan, the financial plan, and corporation and individual financial statements. Many aspects of the loan proposal are incorporated into the business plan, as described in Chapter 2. Lenders will require their own loan applications.

---

does not explain or what has been omitted. Therefore, it is important to use the outline in Table 5.6 as you prepare and present your pro forma.

The building blocks are completed using a conservative, technical approach to ensure that all data are organized and presented to equity and debt participants.

Table 5.7 Sample Cash and Income Statement for JC Penney Rehabilitation
Project (Figures in $)

	Year 1	Year 3	Year 5	Year 7	Year 10
Cash Assumptions					
Equity	50,000				
Loan	50,000				
County subsidy	55,000	40,000	15,000	15,000	15,000
Grants	50,000	100,000	100,000	100,000	100,000
Total Cash	**205,000**	**140,000**	**115,000**	**115,000**	**115,000**
Pro Forma					
Revenues					
Rents	18,980	20,102	21,220	22,347	23,470
Vending machines	786	884	982	1,081	1,179
Fee income	155,000	195,000	235,000	255,000	265,000
Total Revenues	**174,766**	**215,986**	**257,202**	**278,428**	**289,649**
Expenses					
Payroll	140,000	145,000	150,000	155,000	165,000
Taxes	22,550	22,570	22,570	22,650	23,400
Equipment	20,850	6,000	6,000	6,750	6,750
Utilities	11,600	12,000	12,000	12,000	12,000
Legal and accounting	1,288	1,300	1,288	1,288	1,288
Insurance	3,900	3,900	4,300	5,000	5,000
Transportation	12,000	9,000	9,000	9,000	9,000
Professional fees	35,000	15,000	10,000	10,000	10,000
Miscellaneous	25,000	5,000	15,000	15,000	5,000
Total Expenses	**272,188**	**219,770**	**230,158**	**236,688**	**237,438**
Net Operating Income	**(97,422)**	**(3,784)**	**27,044**	**41,740**	**52,211**

Illustration: Brava Valley Community Corporation (BVCC): JC Penney Site

The JC Penney site will be converted into three facilities: a Child Care Center, an Elder Care Center, and a Community Arts Center. A summary of fee income, grant
income, equity, debt, and subsidies was initially prepared. Following the summary, a
10-year pro forma was prepared to document the expected revenues, expenses,
and net operating income only. No yield calculations were made. The board of
directors wanted to determine if the project could provide cash flow. The decision
to review the net operating income was based on the fact that the net operating
income would be the best indication of the net cash available to pay the loan and
taxes on the venture (see Table 5.7).

Although all of the line items in Table 5.7 are estimates only, the BVCC board requested that each of the line items be tested and refined to ensure accuracy for the venture. Their concern over the estimates stemmed from the fee income estimates. The board was not convinced that the fee income forecasts could be achieved.

Discounted Cash Flow Analysis

Economic development projects have been developed with as little as 10% equity and the balance of funds split between conventional loans, public or tax-exempt loans, and a variety of grants or subsidies. Under a multitiered financing approach, layering and allocating specific funds—parking district funds for the parking garage, job training funds for computer training facilities—requires an intensive process. The market dynamics, the size of the facility or service, and the development assumptions must be reevaluated and altered using the pro forma, with the discounted cash flow (DCF) model as the tool.

Earlier in this chapter, we discussed the differing views of value from the private sector, the public sector, and foundations. With investment analysis using the DCF model, we test the financial performance of the project or service, identify financial shortfalls, and evaluate the returns and financial risks associated with the development. Most importantly, we measure value to the private investor or public agency. Because benefits in real estate investments are received over time, the time value of money is important. Cash flows from two investments may be similar in size, but one may return cash to the investor within 2 years and the other in 10 years. The value of the cash flows in 10 years is worth less than the cash flow returning to the investor in 2 years. Because the 10-year cash flow carries with it greater uncertainty, it is expressed in terms of a higher discount rate. The higher the discount rate, the lower the perceived value. The DCF analysis reduces future cash flows to a single point in time so that the investments can be compared and decisions made.

Using the pro forma as the presentation tool and the DCF as the analytical tool, we calculate the requirements for debt and equity based on an overall required return. The investor or private sector decision is more precise and based solely on monetary returns. The public sector will review merely the income and expenses to understand the funding gap that is required to ensure that the development occurs. In a social program rather than a project development, land costs are irrelevant, assuming that fees for services can cover lease payments. Typically, funding will be made on the basis of the performance of the organization.

Investors and lenders in real estate require an overall return that represents at least 7 or 8 percentage points above a "safe rate" or Treasury bill rate. Therefore, if Treasuries were yielding 6%, the *minimum* investor return would yield 13% or

14%. However, today, the investor can invest in mutual funds and receive a variety of returns depending on the fund and the risk. Over the past several years, we have seen annual returns in certain funds as high as 100% and as low as –5%. Mutual funds provide more liquidity and can be more volatile than a development project or a new business enterprise. Therefore, the investor today is typically seeking a return of 20% to 25% for development projects. Investors in technology start-ups often require a minimum of 30% return. Overall, the investor has several options and typically will invest in those instruments or investments that will yield the highest return with the least amount of risk. Thus, the purpose for the DCF analysis is to examine the overall return for a period of time—5 or 10 years—and compare that return to investments in alternative instruments.

We measure the cash inflow (revenues from the project) and outflow (expenses, debt services, and taxes) to arrive at a present worth or present value. The present value measures the future cash flows in today's dollars. The net present value (NPV) is the present value of the sales proceeds and cash flows minus the equity investment.

In preparing the financial plan and testing the scenarios, we build the DCF model, using first the assumptions developed in our market research (see Figure 5.2, Table 5.8). All revenue and expense assumptions are listed under the assumptions, together with project purchase price, equity in the project, loan amount, term, and rate, depreciation schedule, and performance measures.

For the DCF analysis, we will assume that BVCC will retain a for-profit group to develop the BVCC Technology Park. The park will have annual rents ranging from $2.40 per square foot to $3.96 per square foot. Vacancy rates will range from 7% to 10% in the first 5 years and 5% to 6% in Years 6 through 10. Annual revenues from vending machines will be $2,000. The tenants will reimburse the owner for common area maintenance (common area services and utilities, maintenance and security, real estate taxes, and property insurance). Other expenses assumed by the owner will include property management fees (3% of gross revenues), administrative and overhead (4% of gross revenues), miscellaneous expenses (2% of gross revenues), and reserves for repairs (3% of gross revenues). Assumptions for equity, grants and debt (amount, term, and interest rate), tax rates, and depreciation have been developed. See Table 5.8 for the complete model and assumptions.

Figure 5.2 shows the flow of funds. In Part A, we calculate the before-tax cash flow; in Part B, the debt service; in Part C, taxable income; in Part D, the after-tax cash flow; in Part E, sales proceeds; and in Part F, the present value and net present value of proceeds. The key measures are the net present value, the after-tax internal rate of return (IRR) before sale, and the internal rate of return after sale (see Figure 5.2).

Revenues and Expenses

We begin the DCF analysis by examining all sources of gross income. *Gross income* is that income derived from rents or sales, parking, laundry, vending

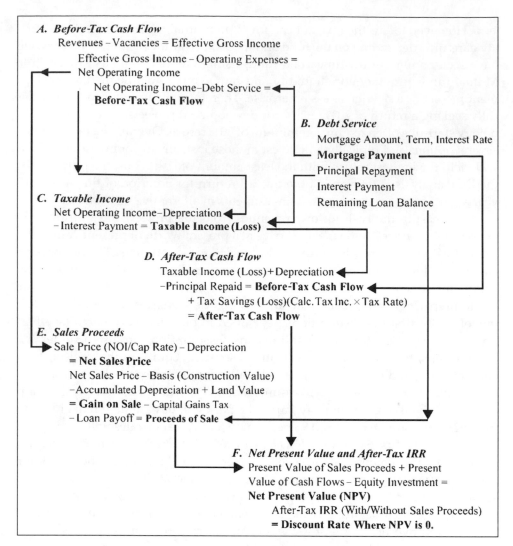

Figure 5.2. Discounted Cash Flow Model Diagram

machines, public telephones, fees for child care, membership fees for exercise studios, and fees and charges for computer and game use in a cybercafe. From the gross income, we subtract the *vacancy allowance and collection loss*. The vacancy allowance and collection loss is typically based on the property history or market trends. The difference between the gross income and vacancy allowance/collection loss is the *effective gross income*.

Let us assume that we have purchased an industrial building (technology center) with 300,000 square feet for $7.5 million. We secured a loan for $6 million from a commercial bank. The bank provided a 15-year loan at 10% interest. The annual average rent for tenants is $5.00 per square foot net of taxes, insurance, maintenance, and common area maintenance (CAM). We assume that 85% of the CAM

(text continues on page 104)

Table 5.8 Commercial/Industrial Pro Forma

A. Assumptions
Revenue Assumptions

Leases	Sq. Ft.	Monthly Rent ($)	Extra Parking Rent ($ Monthly) ($50/Stall)	Monthly Vending Revenues ($)	Monthly Vacancy	Monthly Vacancy, Yrs. 6-10
Building 1: Lt. Manuf.	100,000	40,000		750	7%	5%
Building 2: Lt. Manuf.	75,000	30,000		750	7%	5%
Building 3: Office, R&D	25,000	20,000		0	7%	5%
Building 4: Bulk Warehouse	100,000	35,000		500	7%	5%
Total	**300,000**	**125,000**	**0**	**2,000**	**7%**	**5%**

Expense Assumptions

Property Management: 3.00% of gross revenues

Administrative and Overhead: 4.00% of gross revenues

Common Area Services and Utilities: 6.20% of gross revenues

Maintenance and Security: 6.00% of gross revenues

Real Estate Taxes: 1.02% of purchase price; increase at x% per year—annual, see escalations below

Insurance: 2.00% of gross revenues

Other: 3.00% of gross revenues

Reserves for Repairs: 3.00% of gross revenues

Revenue Escalation Options: 2.00% or 3.00%

(continued)

Table 5.8 Continued

Expense Escalation Options: 3.00% or 4.00%

Property Tax Escalations: 2.00%

CAM Billbacks

   Year 1: $246,075

   Year 2: $252,807

   Year 3: $259,728

   Year 4: $266,843

   Year 5: $274,159

   Year 6: $281,679

Purchase Price: $7,500,000

Other Assumptions

Equity in Project: $1,500,000

Loan Amount: $6,000,000 (You calculate based on project costs less equity)

Term (Years): 15

Rate of Loan: 10%

Remaining Loan Balance (Principal Amount Less Principal Repaid)

Interest Paid During Period (IPMT)

Principal Repaid (PPMT)

Mortgage Payment (PMT)

Depreciation (Improvement/Useful Life): Commercial = 39 years

Income Tax Rate: 31%

Capital Gains Tax Rate: 20%

Percent Value of Land: 20%

Percent Value of Center: 80%

Sales Commission: 6%

Performance Measures

Debt Coverage Ratio: 1.25-1.50+

Capitalization Rate Options: 9%, 15%, 12%, or 14%

Desired IRR: 15%-20%

Discount Rate: 10%-15%

Cumulative Depreciation Calculations: (Purchase Price × % Improvement/Useful Life)

Year 1: $153,846
Year 2: $307,692
Year 3: $461,538
Year 4: $615,385
Year 5: $769,231
Year 6: $923,077

(continued)

Table 5.8 Continued

B. Before-Tax Cash Flow (in $, Except for Debt Coverage Ratio)

	Year 0	Year 1	Year 2	Year 3	Year 4	Year 5	Year 6
End-of-Year Equity Investment	(1,500,000)						
Revenues							
Gross Revenues		1,500,000	1,530,000	1,560,600	1,591,812	1,623,648	1,656,121
Other Income		24,000	24,480	24,970	25,469	25,978	26,498
Less: Vacancy and Credit Loss		105,000	107,100	109,242	111,427	113,655	82,806
Add: CAM Billbacks		246,075	252,807	259,728	266,843	274,159	281,679
Effective Gross Income		**1,665,075**	**1,700,187**	**1,736,056**	**1,772,697**	**1,810,130**	**1,881,493**
Operating Expenses							
Property Management		45,000	46,350	47,741	49,173	50,648	52,167
Administrative and Overhead		60,000	61,800	63,654	65,564	67,531	69,556
Common Area Utilities and Services		93,000	95,790	98,664	101,624	104,672	107,812
Maintenance and Security		90,000	92,700	95,481	98,345	101,296	104,335
Real Estate Taxes		76,500	78,030	79,591	81,182	82,806	84,462
Insurance		30,000	30,900	31,827	32,782	33,765	34,778
Other		45,000	46,350	47,741	49,173	50,648	52,167
Reserves for Repairs		45,000	46,350	47,741	49,173	50,648	52,167
Total Expenses		**484,500**	**498,270**	**512,438**	**527,015**	**542,014**	**557,446**
Net Operating Income		**1,180,575**	**1,201,917**	**1,223,618**	**1,245,682**	**1,268,116**	**1,324,047**
Debt Service		773,716	773,716	773,716	773,716	773,716	773,716
Debt Coverage Ratio		1.53	1.55	1.58	1.61	1.64	1.71
Cash Flow Before Taxes		**406,859**	**428,201**	**449,902**	**471,967**	**494,400**	**550,331**

C. Calculation of Debt Service (End-of-Year Calculation, in $)

	Year 1	Year 2	Year 3	Year 4	Year 5	Year 6
Mortgage Payment	773,716	773,716	773,716	773,716	773,716	773,716
Interest Paid During Period	583,396	563,467	541,451	517,130	490,262	460,581
Principal Repaid During Period	190,320	210,249	232,265	256,586	283,454	313,135
Remaining Loan Balance	5,809,680	5,599,431	$5,367,166	5,110,580	4,827,126	4,513,991

D. Calculation of Taxable Income (in $)

	Year 1	Year 2	Year 3	Year 4	Year 5	Year 6
Net Operating Income	1,180,575	1,201,917	1,223,618	1,245,682	1,268,116	1,324,047
Less: Depreciation	(153,846)	(153,846)	(153,846)	(153,846)	(153,846)	(153,846)
Less: Interest (To Be Calc.)	(583,396)	(563,467)	(541,451)	(517,130)	(490,262)	(460,581)
Taxable Income (Loss)	443,333	484,604	528,321	574,707	624,008	709,620

(continued)

Table 5.8 Continued

E. After-Tax Cash Flow (in $)

	Year 1	Year 2	Year 3	Year 4	Year 5	Year 6
Taxable Income (or Loss)	443,333	484,604	528,321	574,707	624,008	709,620
Plus: Depreciation	153,846	153,846	153,846	153,846	153,846	153,846
Less: Principal Repaid (To Be Calc.)	(190,320)	(210,249)	(232,265)	(256,586)	(283,454)	(313,135)
Cash Flow Before Tax	406,859	428,201	449,902	471,967	494,400	550,331
Plus: Tax Savings (or Tax) (Calc. Tax. Inc. × Tax Rate)	(137,433)	(150,227)	(163,779)	(178,159)	(193,443)	(219,982)
Cash Flow After Tax	**269,426**	**277,974**	**286,123**	**293,808**	**300,958**	**330,349**

F. Calculation of Sales Proceeds (in $)

	Year 1	Year 2	Year 3	Year 4	Year 5	Year 6
Sales Price (NOI/Cap Rate)	8,012,780	8,157,452	8,304,549	8,454,107	8,826,977	8,826,977
Less: Selling Expenses	480,767	489,447	498,273	507,246	529,619	529,619
Net Sales Price	7,532,013	7,668,005	7,806,276	7,946,861	8,297,358	8,297,358
Less: Basis (Basis: Construction Value Less Accumulated Depreciation Plus Land Value)	7,346,154	7,192,308	7,038,462	6,884,615	6,730,769	6,730,769
Gain on Sale	185,859	475,697	767,815	1,062,246	1,566,589	1,566,589
Less: Capital Gains Tax	37,172	95,139	153,563	212,449	313,318	313,318
Less: Loan Payoff (To Be Calc.)	5,809,680	5,599,431	5,367,166	5,110,580	4,827,126	4,513,991
Proceeds of Sale	**1,685,161**	**1,973,434**	**2,285,547**	**2,623,832**	**3,156,914**	**3,470,049**

G. Calculation of Present Value and Net Present Value of Proceeds

	Year 0	Year 1	Year 2	Year 3	Year 4	Year 5	Year 6
Proceeds of Sale		$1,685,161	$1,973,434	$2,285,547	$2,623,832	$3,156,914	$3,470,049
Present Value Factor, 1-5 Yrs., 15%		0.8696	0.7561	0.6575	0.5718	0.4972	0.4323
Present Value of Sale Proceeds		$1,465,416	$1,492,114	$1,502,747	$1,500,307	$1,569,618	$1,500,102
Cash Flow After Tax	($1,500,000)	$269,426	$277,974	$286,123	$293,808	$300,958	$330,349
Present Value Factors (Discount Rate From Investors)	0.15	0.8696	0.7561	0.6575	0.5718	0.4972	0.4323
Present Value of Cash Flow After Tax and Equity Investment	($1,500,000)	$234,293	$210,176	$188,126	$167,999	$1,49 636	$142,810
Net Present Value After Tax (Calculation Based on PV of Sales Proceeds Plus PV of Cash Flows Less Equity Investment)		$226,406	$447,216	$632,902	$870,212	$950,332	
Internal Rates of Return							
Cash-on-Cash Return		18%	19%	19%	20%	20%	22%
After-Tax IRR (Not Including Sales Proceeds)		a	a	−24.29%	−10.39%	−1.58%	
Cash Flow After Tax							
Year 1	($1,500,000)	$1,954,587					
Year 2	($1,500,000)	$269,426	$2,251,408				
Year 3	($1,500,000)	$269,426	$277,974	$2,571,670			
Year 4	($1,500,000)	$269,426	$277,974	$286,123	$2,917,639		
Year 5	($1,500,000)	$269,426	$277,974	$286,123	$293,808	$1,986,119	
After-Tax IRR (Including Sales)		30%	32%	31%	31%	20%	

a. Return cannot be measured.

charges are allocated and billed back to the tenants and thus become a line item ("billbacks") on the revenue side of the pro forma. The gross revenues in the first year will be $1,500,000. Other income from vending machines will be $24,000. Billbacks for CAM are $246,075. If the vacancy allowance and collection loss is 7%, the effective gross income will be $1,665,075.

Operating expenses include management fees, payroll, repairs, utilities, accounting and legal fees, marketing and advertising, property taxes, and insurance premiums. The operating expenses are location and property type comparable and typically range from 30% to 50% of gross income depending on property type. These costs can be estimated and confirmed through property managers and real estate brokers in the market. The operating costs should be precise. If we continue with our industrial building example, operating expenses will be 32.3% of gross revenues, or $484,500 in the first year.

Included in our expenses are major repairs and reserves for replacement. They are deducted from the effective gross income along with the operating expenses. Major repairs and replacements are typically categorized as "reserves," with funds set aside and used to replace plumbing systems; heating, ventilation, and air conditioning systems; roofs; or major appliances or equipment. They are usually calculated on the basis of a percentage of gross income—2% to 3%— depending on the age of the building. Continuing with our example, major repairs and reserves will be 3%, or $45,000 annually. Once again, the number is included in operating expenses.

Inflation will affect both revenues and expenses and should be adjusted annually to reflect local inflation factors. For example, research and development space in one part of the United States will escalate at a different rate from another area, based on the overall composition of the economic base and the availability of space.

▬ Net Operating Income and Before-Tax Cash Flow

The difference between the effective gross income plus billbacks and the total expenses is the *net operating income*. Using our industrial building example, net operating income would be calculated as follows: gross revenues ($1,500,000) plus other income ($24,000) plus billbacks ($246,075) less vacancy and collection loss (–$150,000) equals effective gross income ($1,665,075). Effective gross income ($1,665,075) less operating expenses and reserves ($484,500) equals net operating income ($1,180,575).

Subtracting the annual debt service ($773,716) from the net operating income will yield the *cash flow before taxes*. Our industrial building will yield $406,859 before taxes.

Net operating income is the realizable revenue. It is also a measure used by conventional lenders to assess the ability of the project team to make loan payments over a period of years. Lenders typically view an annual *debt coverage ratio*

Table 5.9 Net Operating Income and Before-Tax Cash Flow Formulas

Gross Revenues + Other Income – Vacancy and Credit Loss + Billbacks =
    Effective Gross Income

Effective Gross Income – Operating Expenses = Net Operating Income

Net Operating Income – Debt Service = Before-Tax Cash Flow

Table 5.10 Break-Even Ratio (in $000s and Rounded)

	Year 1	Year 2	Year 3	Year 4	Year 5
Formula example	$1.26/$1.50	$1.27/$1.53	$1.29/$1.56	$1.30/$1.60	$1.32/$1.62
Break-even ratio	83.9%	83.1%	82.4%	81.7%	81.0%

(annual net operating income divided by annual debt service payments) of 1.25 to
1.50 in the initial year of the project to be acceptable. Debt coverage ratios should
increase over time, signaling increases in effective gross income or reduction in
expenses and growth in net operating income. Using our example, the debt coverage ratio equals 1.53, or a level that is acceptable to lenders (Table 5.9).

Lenders also measure the *break-even ratio,* a measure of financial risk. The
break-even analysis allows the lender to calculate the "break-even" occupancy
level at a projected rent level to determine if there is sufficient revenue to cover
operating expenses and debt service. The lower the break-even ratio, the more
financially stable the project; the higher the break-even ratio, the higher the financial risk. Break-even ratios typically range from 75% to 90%, with the higher percentage in the early years of the project. Break-even ratios will decline over time as
revenue increases. Increasing both revenues and expenses in our industrial building example, we find that the break-even ratio analysis indicates a positive trend
downward from 83.9% in the first year to 81% in the fifth year. The analysis shows
that revenues are sufficient to cover both operating expenses (plus annual
increases) and debt service to the extent that most lenders would be comfortable
with the risk they would assume (Table 5.10).

▬ After-Tax Cash Flow

Many investors may decide not to continue the analysis at this point, being
secure with the cash flow before taxes. However, the more sophisticated investors,
lenders, and Wall Street firms will proceed with the cash flow after-tax analysis
and the determination of the internal rates of return with and without sale

Table 5.11 After-Tax Cash Flow (in $)

Taxable Income (or Loss)	443,333
Plus: Depreciation	153,846
Less: Principal to Be Repaid	(190,320)
Cash Flow Before Tax	**406,859**
Plus Tax Savings (or Tax)	(137,433)
Cash Flow After Tax	**269,426**

proceeds. The internal rate of return, or the yield to the investor, is the single measure used to compare alternative investments.

To calculate the *cash flow after taxes,* we simple deduct any tax liability or add any tax savings to the cash flow. If there is no tax liability, the cash flow is considered sheltered. Tax savings are an important gift in real estate investment and come primarily through depreciation and interest deduction from mortgage payments. Depreciation is a deductible expense for investment or commercial property, although it is not an actual dollar expense. Only buildings are depreciated. There is no depreciation for the land component. Depreciation reflects the inherent wear and tear of the property over the holding period. Thus, from the net operating income, we deduct depreciation and interest payments to arrive at the taxable income. As Table 5.11 indicates, we take the taxable income, add back the depreciation, and subtract the principal repaid to arrive at the cash flow before taxes. After applying the appropriate tax rate, we derive the tax savings or tax that is subtracted or added to the cash flow before taxes. The net result is the cash flow after taxes.

▬ Sale Proceeds

As indicated in the DCF model presented in Table 5.8, a series of calculations occurs to arrive at the proceeds of sale. The most frequently used approach is to capitalize the net operating income, reflecting the typical sale of investment properties. Capitalization rates are derived from the marketplace. Differing property types carry with them different capitalization rates, and rates vary from year to year. The higher the capitalization rate, the lower the sale proceeds. The use of the capitalization technique occurs on the income following the year the property is sold. Sale and marketing expenses are deducted from the capitalized sale price to yield the net sales price. Deducting the basis (construction value less accumulated depreciation plus land value) will then provide the gain on sale to which the capital gains tax is deducted along with the loan payoff. The net number is our sales proceeds (Table 5.12).

Table 5.12 Sales Proceeds Formula

Sale Price = Net Operating Income/Capitalization Rate
Basis = (Construction Value – Accumulated Depreciation) + Land Value
Sale Price – Selling Expenses = Net Sales Price
Net Sales Price – Basis = Gain on Sale
Gain on Sale – Capital Gains Tax – Loan Payoff = Proceeds of Sale

▬ After-Tax Net Present Value and Internal Rate of Return

The net present value of the investment is derived by measuring the after-tax cash flows annually during the holding period. The net present value is the sum of the discounted cash flows and sales proceeds. If the sum of the expected discounted cash flows exceeds the purchase price or cost of the investment, the likely decision will be to invest in the project. If the sum is less than the purchase price or cost of the investment, the likely decision will be to withdraw the purchase offer or negotiate a lower purchase price.

The most common method of investment return analysis is measuring yield or the internal rate of return (IRR). The dollar flows from the project over time are compared to the equity investment. The IRR is the discount rate at which the net present value is zero. It is the yield to the investor. If the IRR is less than the desired rate of return, the investor will choose to reject the investment; if it is equal to or greater than the desired return rate, the investor will typically invest.

The DCF model is used to evaluate risk and reward in projects. Because numerous variables affect the returns of the project, the reader is encouraged to perform sensitivity analyses for income, expenses, inflation factors, purchase prices, equity, and debt to understand fully the impact on the net present value and IRR. Gap financing, subsidies, and grants can all be used to write down the costs, thereby turning an unprofitable project to a profitable venture.

Evaluating Risk

Because economic development projects often carry inherent market and financial risks, layering of funds spreads the risk so that one financial institution does not carry the full burden. In financing terms, we consider the pooling of funds for the development or service as "syndication" of the financing—each financial entity will be at risk for the amount of money invested in the project. The risk will be minimized through guarantees by the developers and government agencies.

Risk mitigation strategies are commonly used by private and conventional lenders (see Table 5.2) but are often pushed aside by the public sector and even the developer. Public agencies see certain catalyst projects and more often large-scale development as enormous opportunities to generate tax revenues and improve urban centers. These agencies often neglect reviewing and developing risk mitigation strategies to ensure that the "ideal" projects do not fail. More often, we see cities and counties becoming liable for bond debt when a project does not generate sufficient revenues or expenses are not controlled. Developers often refuse to address risk measures that protect not only them but also their lenders. Outlining potential risks and revising the financial model to a more conservative approach will often provide greater credibility for the developer when he or she is seeking funding. Risks for both the private and public sectors can be reduced further through (a) the performance of sensitivity analysis to identify how the project will perform in a poor economy and a prosperous economy, (b) understanding of borrower and project risk, and (c) the development of a mitigation strategy. Types of risks and mitigation strategies for real estate projects are outlined in Table 5.13.

The most critical components of the financial plan are the risk analysis and testing assumptions for sensitivity to the NPV and IRR. The sensitivity and risk analyses in the early planning stages will help prevent failed projects, loan defaults, and bankruptcies. All parties to a transaction should perform these analyses to ensure that their financial interests are protected.

Requirements for a Profitable Project

What are the basic ingredients for a profitable project? The private sector will maintain that high returns in excess of 20% are necessary. The public sector will agree that a completed and fully leased building serving to revitalize a neighborhood or urban downtown and generating revenues is profitable. The lenders will define a profitable project as one that provides sufficient cash coverage of the debt service through a high revenue stream and controlled expenses. The project or service will cover the debt so that mortgage and other loan payments can be made. All are correct; however, more is required. A profitable project will have the following characteristics:

- There is market demand to support the project, as evidenced by a demand by prospective tenants for space.
- The project receives necessary approvals from the city or county.
- Approvals are granted within a reasonable period of time.
- All environmental issues are resolved, and the lenders are in agreement with the resolution.

Table 5.13 Risks and Mitigation Strategies for Loans

Risk	*Mitigation Strategy*
Borrower Risk	
Track record	Review financial performance for each project or business; conduct interview process, check references, and seek approvals from independent parties.
Current capacity	Verify development or business schedule; check other projects or ventures.
Financial and organizational status of borrower	Review corporate and individual financial statements and tax returns for 3 years. Require equity; require credit enhancement if financial statement does not meet high standards.
Project Risk	
Failure or inability to complete project	
Unexpected physical issues/problems	Review construction costs, plans, and schedule; secure independent construction manager to monitor; require environmental assessment for development projects.
Cost overruns	Retain independent construction or operations manager to monitor construction or operations draws; establish contingency; negotiate maximum soft cost amount; negotiate who will be responsible for cost overruns.
General contractor performance	Require owner/developer equity investment; require surety bonds and letters of credit for performance completion; require completion guarantees and define penalties for failure to complete; retain a percentage of the total contract and developer profit until project is completed.
Financial failure	Secure completion guarantees; obtain letter of credit and performance bonds; retain percentage of payments to developer. For businesses, secure personal or other guarantees, letter of credit, and reserves.
Reduction in property/business value	Require additional collateral if property or business value declines below required debt/equity ratio.
Loss of public subsidy, grants	Obtain guarantees from public agency, foundations, and corporations.
Loan Default	
Financial mismanagement by developer and property manager or business officers	Obtain developer or executive equity; obtain payment guarantees; retain experienced property managers and executives and check references.
Market decline	Obtain annual appraisals; review market conditions to ensure that loan-to-value ratio does not exceed underwriting standards.
Poor systems	Review management systems to verify that revenues and expenses or rents and expenses can be monitored; review monthly statements and lease renews for appropriate escalation of rents and controls on expenses; require operating reserve to cover costs if necessary.

- The developer or business group contributes to the equity investment and provides a letter of credit to guarantee any construction or business short-falls.
- Debt is provided by a group of lenders offering a range of interest rates to reduce the overall costs.
- Debt coverage ratio in the first 5 years is expected to exceed 2.5.
- IRRs without a sale exceed 15% by the fifth year.
- The developers or business executives have a strong track record showing profitable ventures, business completions, and/or project completions on schedule.
- For development projects, the tenants have strong credit.
- The property manager is experienced and proactive with the tenants.

Case Study 4: The BVCC Venture Financial Center

The BVCC board has concluded that it will negotiate with the Air Force base. After several studies and technical assistance reports, with assistance from the county, the board has decided on a new approach to rehabilitate part of the base separate from the technology park. The board is examining a set of enterprises to take up part of the vacant space on the base. These enterprises are outlined below, along with requirements for square footage, rents per square foot, and operations entities.

Projects	Sq. Ft.	Rent/Sq. Ft. (in $)	Operation Entity
Cyber center	1,200	0.75	Nampo Tribe
Community library	2,300	1.80	Brava County
Copy center	1,000	2.00	Private lease
Mail stop	500	0.75	Private lease
Employment counseling	750	2.00	State of N.M.
Health club	2,000	1.50	Private lease
Virtual golf	1,500	1.50	Private lease
Video store	750	1.25	Private lease
Convenience store	1,200	1.25	Private lease
Fast food	500	1.25	Private lease
Book store	750	1.25	Private lease
Resale shop	500	1.50	BVCC Mothers
Juice bar	750	0.75	BVCC Youth

Using the DCF model in Table 5.8 or accessing the computer model at www. sagepub.com/giles, prepare responses to Questions 1 and 2.

You have decided to go forward to develop part of the Air Force base. Using your preferred alternative, prepare the financial plan as outlined below.

1. Based on the market assumptions used in Chapter 4, prepare a financial plan for the base. Director Mohamid-Gonzales is seeking financing for this project from foundations, the state through the state enterprise zone strategy, the county local development corporation, and a local bank. As part of the Reuse Plan, the entire base has been sold to BVCC for $1.00. The actual market value of the base is $1.5 million, based on a recent appraisal. The facilities scheduled to be converted require rehabilitation. Part of the base will be used as a technology park, as stated earlier in this chapter. BVCC will maintain the base, including roads and infrastructure. Private security will be required for the entire base.

2. Outline the assumptions used for three scenarios—best case, most likely case, and worst case. Include the following sections:

   - Project Description: What is the project?
   - Organizational Strategy: Who will develop the project, and how is the organization structured?
   - Financing Scenarios (three scenarios)
   - Financial Plan
   - Sources and Uses of Funds
   - Project Income
   - Project Cash Flow
   - Profit
   - Before-Tax IRR, With and Without a Sale at the End of the Holding Period
   - Recommendations

Case Exercises

1. DCF Model

   a. Using the DCF model, vary the inflation factors for revenues and expenses on an annual basis. What effects do these variations have on the before-tax cash flow?

   b. Increase the loan amount 5%. What is the effect on the IRR?

   c. Reduce the land costs through the use of grants. List the grants that you use, and compare the IRR before and after the reduction of land costs.

   d. If you choose to sell the property in Year 2 rather than Year 5, what is the effect on the IRR?

2. Loan Proposal

   a. Prepare an outline for a loan proposal covering the following items:
      - Summary

- Business plan
- Financial plan
- Corporation financial statement
- Other personal data

b. List the major components of your project and why you believe that your project should be financed. What are the benefits and risks associated with the project?

3. Matching Funds With Projects

You have been given a series of real estate and business development transactions. Using the list of funding sources below, match the appropriate funds with the transactions. If appropriate, you may use more than one for a transaction.

a. Industrial development bonds

b. Tax-exempt bonds for housing

c. Venture network for start-up technology companies

d. Commercial bank

e. Business loan—economic development bank

a. $2 million loan for the acquisition and development of the technology industrial park _____

b. $1.5 million loan for the development of Interstate Apartments as a senior housing project _____

c. $500,000 in interim financing _____

d. Start-up software company with strong business plan _____

e. $850,000 for expansion of a Home Depot franchise in the empowerment zone that will create 15 new jobs _____

Case Study 5: Community Apartments or Tech Center Financial Plan

Using the DCF model in Table 5.6 or accessing the computer model on the Web site, prepare the responses to Questions 1 and 2 as in the above exercise.

You have decided to recommend either an apartment project or a "tech center" on the industrial-zoned land. Using your preferred alternative, prepare the financial plan as indicated below.

1. Outline the assumptions used for three scenarios—best case, least likely case, and most likely case.

2. Include the following sections:

- Project Description: What is the project?

- Organizational Strategy: Who will develop the project, and how is the organization structured?
- Financing Scenarios (three scenarios)
- Financial Plan
- Sources and Uses of Funds
- Project Income
- Project Cash Flow
- Profit
- Before-Tax IRR, With and Without a Sale at the End of the Holding Period
- Recommendations

Case Exercises

1. DCF Model
   a. Using the DCF model, vary the inflation factors for revenues and expenses on an annual basis. What effects do these variations have on the before-tax cash flow?
   b. Increase the loan amount 5%. What is the effect on the IRR?
   c. Reduce the land costs through the use of grants. List the grants that you use, and compare the IRR before and after the reduction of land costs.
   d. If you choose to sell the property in Year 2 rather than Year 5, what is the effect on the IRR?
2. Loan Proposal
   a. Prepare an outline for a loan proposal covering the following items:
      - Summary
      - Business plan
      - Financial plan
      - Corporation financial statement
      - Other personal data
   b. List the major components of your project and why you believe that your project should be financed. What are the benefits and risks associated with the project?

6
CHAPTER

Accessing the Money for Local Economic Development

L ocating and accessing money for a local economic development program is difficult. Those who have the money to invest or lend have strict hurdles that must be overcome by corporations, organizations, or community groups seeking funding. Although the hurdles can be overcome by new economy or old economy corporations, local economic development corporations must demonstrate financial and organizational strength often lacking in these ventures. Before these money sources become an equity partner or a creditor, they will want details on the product or program, business and personal financial statements, loan guarantees, and much more.

Accessing the money requires a sound corporate strategy for the organization as well as the project. Foundations, corporations, and government agencies are seldom interested in paying for the basic overhead to maintain the organization. They will often assist with funding the project but not with funding the start-up local economic development company. Therefore, a strategy to access money for the entire venture—the company and its program—is needed.

This chapter is divided into three sections:

- The strategy to access funds
- The sources of equity, debt, and gap funding
- The components of a proposal

115

The Strategy to Access Funds

The strategy to access funds will depend on the local economic development program. Development projects will have access to specific funding sources from commercial banks, insurance companies, and Wall Street, among others. Funding for products and services may require more targeting of institutions. Nonprofit organizations have other agencies and foundations that will support the program if the need can be established. The following steps have been used successfully by many organizations.

▬ Identify Money Required

The first step in identifying the money required to build an organization and develop a project is to prepare a budget. Budget preparation is a two-part process: (a) analyzing and documenting start-up expenses plus revenues and operating expenses for the organization and (b) performing a detailed cost analysis and budget to develop or construct the program, project, or service. Both the organization budget and project budget should cover a 5-year time period. The rationale for breaking the overall budget into two components is that funding sources may be different for each component. For example, foundations may provide the seed money to build an organization that will deliver child care services. The bank, however, may lend only on the facility or the child care operations—not on the organization operations. Banks typically require corporate stability and a track record from business professionals.

▬ Identify Sources to Access Money

The type of money the organization needs and what sources it may access are important. If funding is needed initially for the basic operation of the business, then raising money from within the community, local churches, or individual donors may be a top priority. If money is needed for a specific project, a broad base of support may be obtained from local, state, federal, and institutional sources. For example, the development of a technology center with job training for the local workforce will often draw the broad support needed to access funds from many sources.

▬ Choose a Strong Financial Staff

Most organizations expect the CEO, CFO, or executive director to raise funds. However, executives do not know all the people required to access money.

Furthermore, it takes time to cultivate and motivate investors and lenders. Professionals experienced in accessing public, private, and venture funds are essential. Staff from the organization may be used to support these outside professionals. A local United Way will have a variety of fundraisers to access funds from local businesses, philanthropists, and residents. The organization typically has the CEO, CFO, and board members seeking contributions from the local business community. In addition, the organization has a fund-raising professional organizing the campaign and retaining a paid campaign staff. The volunteers associated with the annual giving include the board of directors and key business leaders who donate their time to access the corporate network.

— Prioritize Funding Sources

Organizational structures enhance or limit the ability of a group to raise money. If the organization is a nonprofit, it will have certain doors open to it; if the organization is a real estate development group, it will have a broader group of funding sources available; if the organization is a technology company, specific venture funds exist to support new companies and products. It is important for the organization to develop a very clear priority list of funding sources.

The funding sources must match the type of recipient organization required by the source. For example, certain foundations or corporations will grant money only to nonprofits. Others, including federal, state, and local agencies, will offer funds if the project targets specific income or ethnic groups. For example, the U.S. Department of Housing and Urban Development (HUD) Safe Neighborhood Grants are targeted to managers of public or privately owned multifamily housing for low-income groups. The funds are used to pay for additional local police presence and other security measures to reduce crime in and around the housing. A list of foundations and grantsmakers can be found on the following Web sites: www.philanthropy.com; www.hud.gov; www.foundationcenter.org; www.webcrawler.com (search grantsnet and foundations for specific data).

— Select the Funding Mix

The organization must establish guidelines for the layering of funds from different sources. Lenders and investors often have requirements regarding subordination of loans, requirements for the number of investors and priority calls on returns on equity investment, and the overall mix of debt and equity. Government regulations with accompanying legal penalties can determine the direction and nature of an entire organization. The same is also true if funds are secured from venture capitalists. For example, a small federal contribution to a child care center brings with it an enormous amount of paperwork to justify every hiring decision, the type of food served at the center, and the quality of the facility. Simple corporate contributions may require only that the group meet applicable local laws. Funding secured from venture capitalists for a start-up technology enterprises carries with it an understanding that between 10% and 30% of equity ownership may be given directly to the venture capitalist, resulting in a dilution of ownership for the return of much-needed capital.

Funding Sources

The type of funding we seek for programs, projects, or services is generally categorized into three classes: equity, debt, and gap funding in the form of grants or low-interest loans. A special category of corporate and government funding can be used as equity, debt, or gap funding.

▬ Equity

Money for the initial start-up of an enterprise typically comes from "home." Many new businesses and community projects obtain their initial funding from relatives and friends of the founders of the organization. No matter whether the project is a small business or a community senior center, some funding must come from local sources. In the case of the senior center, the initial capital outlay may come from a local church that donates its building or a portion of the weekly Sunday collection for a year to open the center for its senior citizens. In the case of a neighborhood business, funding often comes from the entrepreneur's savings, from relatives, or from friends.

Venture capital in larger amounts is available through formal sources. In the past several decades, a network of venture capital firms has emerged. These firms finance almost everything from technology companies to home businesses. A comprehensive list of venture capitalists is available through local and regional venture networks and the Internet. The business journals around the United States list the top venture capital firms. The information can be found at www.amcity.com.

Wall Street investment banks are increasingly venture fund sources. Backed by securitized financial instruments traded openly on the stock market, these firms have funds available for many types of projects. They look for projects that will make profits and executives who can build organizations so that the company may engage in initial public offerings (IPOs).

Well-documented strategic business plans provide the written tool that venture capitalists and Wall Street firms require before investment. The strategic business plans are mandatory and will often be the first step before a funding request. The plans will be reviewed by analysts, and the financial statements will be compared to entry requirements established by the venture capitalists or Wall Street firms.

Community or social investment firms are also active in various areas of the country. These organizations, such as local initiative support corporations (LISCs), act as intermediaries for other capital institutions like banks and insurance companies and even mutual funds that look for socially responsible investments generating modest returns. These organizations operate in the same manner as the larger investment organizations; however, their expectations on the investment return are smaller.

Though most of these organizations expect to secure a return on their investment at least equal to the rate of return on a government bond, they will generally assume a larger risk to meet social goals. These goals may include but are not limited to building affordable housing for low-income workers, providing a senior center, or providing a day care center within an urban center.

Small business investment corporations (SBICs) are government-backed investment organizations that are locally based. These organizations have Small Business Administration (SBA) backing. The SBIC raises funds locally for small business investment, and the SBA backs this with a dollar-for-dollar investment pools. As a result, the SBIC has strong leverage for its funding.

SBICs invest in local small business start-ups that usually encounter enormous hurdles at the outset. These hurdles, for example, may include business location or the owner's lack of experience. SBICs can invest in anything ranging from technology companies to restaurants serving a low-income neighborhood.

They act as venture capital firms and expect a reasonable return on the investments. SBICs must have over $300,000 in opening funds. Starting an SBIC requires securing these opening funds from other sources. The initial $300,000 represents a considerable hurdle for many low-income communities; however, through grants from corporate donations and/or grants from larger foundations, the initial $300,000 hurdle may be overcome.

▬ Private Debt Sources

Banks, insurance companies, and credit unions will lend to new start-up businesses at reasonable interest rates. Banks make their money available in two ways. First, they provide secured loans. That is, they will loan money to a new enterprise if the owner will offer as collateral (e.g., a guarantee to pay) some other asset equal to the value of the loan. In many instances, the owner has to put up his or her home or car, or a community agency is required to take a second mortgage against one of its more successful housing projects.

The other form of bank lending for small businesses is through lines of credit. In this form of lending, the bank estimates the amount of income that the organization will generate and makes periodic small loans to the organization up to some maximum amount. The organization, in turn, makes regular payments on the loan in a manner similar to a credit card payment. In fact, large lines of credit on consumer credit cards are replacing lines of credit for many of the nation's small businesses.

Insurance companies are paid through policyholder premiums. To pay for future benefits, insurance companies invest the funds they receive. As a result, they invest in real estate, stocks, and other entities. Insurance companies seldom invest directly in a small or large corporation. They will, however, invest in banks and investment companies that make investments to small or large corporations.

Insurance money is typically long-term debt at generally favorable interest rates. Finally, insurance companies are becoming increasingly more sensitive to neighborhood needs. These companies are investing in local intermediaries such as LISCs and community development banks as a means to funnel their money in neighborhood or community projects.

— Government Funding

Government money comes in three forms—loans, grants, and tax credits. Loans are used as debt, grants are used to offset or reduce debt, and tax credits are used as incentives for equity investment.

Loans

Loans differ from grants in only one respect, *repayment.* Frequently, the same agency that offers grants offers loans for the same programs. Housing is the best example. A group can obtain a block grant allocation for low-income or senior housing needs assessment and design and a government loan for construction. In many instances, the grants and loans can be allocated at the same time from the same government agency. But this is not always the case. It is possible to get a grant for senior housing from HUD and a loan for the health care component from the U.S. Department of Health and Human Services. The guidelines and accounting procedures for each may be drastically different. If a program-related investment is received from a foundation, the agency may have still another set of books to keep.

Government loans are made primarily for community facilities and infrastructure where user fees, taxes, and/or levees will be the sources of repayment for the loans. These loans must be repaid to the U.S. Treasury at specified intervals. Usually, they have very favorable terms and various forgiveness clauses. In some cases, the loans may be indirect through a government loan guarantee. The state or federal government guarantees the repayment of a loan to a commercial bank. The bank must exercise nominal prudence in advancing the loan, but the ultimate repayment is guaranteed by the government.

Rural areas have special sets of loans and grants to provide housing and infrastructure ranging from telephones and sewers to housing. Each state has a rural development program that provides a mixture of state and federal funding to promote rural modernization.

In urban areas, government loans are typically available for large-scale urban projects such as downtown revitalization. Federal loans are used by localities and combined with private capital to develop inner-city projects. These projects must demonstrate a means of repayment. The business plan is the means by which these lenders can evaluate the sources of loan repayment.

Grants

Grants are outright gifts typically used to reduce the amount of debt required by the organization. A grant requires that the recipient fulfill the obligations set forth in the grant request. There is no expectation of repayment.

Formula or Block Grants. Formula or block grants are funds given to governmental units based on the allocation of formulas such as population or level of poverty in a local area. The grants are used for specified purposes. State governments are the primary direct beneficiaries of block and formula grants. The federal government will allocate block or formula grants to states with certain restrictions to conduct certain activities that are deemed to be in the national interest. Health care and roads are examples. The state may reallocate the grants to state bureaucracies to administer the funds. An example is the case of road improvement funds allocated to the state transportation agency for funding local road construction or repairs.

Each state has a different mechanism for the allocation of block grants. In a few instances, the block grant goes directly to cities or counties, which in turn fund locally based programs or conduct the activity directly. No matter what the venue, community groups receive portions of block grants and are bound by the terms of the grant.

The most well-known and visible block grants are for housing, crime prevention, and infrastructure. Community groups who work in these fields know the grant systems well. In many instances, block grants are allocated to local groups via some form of competition. Community groups who want to compete need to know the rules so they can write proposals to meet all of the federal legal requirements.

Project Grants. Project grants are made to focus money toward specific projects. This is a change from the past, when broad project funding was available. Although the goal has been to eliminate direct funding, it is unlikely that the federal government or states will eliminate all special program areas or projects for grants.

Program grants are the best means of marshaling public attention and private resources for a significant set of problems. In some areas such as the arts, a national agency, the National Endowment for the Arts, has a stake in project grant making. In other fields, the federal government or state government shapes the agenda without delivering the service directly. Cancer and AIDS research are cases in point where the federal and state initiatives have shaped funding from the private sector as well.

Community-based grants in this area take a variety of forms. In some cases, such as crime prevention, special grants are made available to communities that use community policing. Community policing may well involve local community-based organizations. Similar grants are available in employment, training, and welfare to support community programs promoting the transition of welfare recipients into the workforce.

Empowerment and enterprise zones are federal and state government grants delivered to impoverished areas. The grants are targeted to improve local infra-structure, revive commerce, and stimulate the local job base. Most states have some form of enterprise zone or area where a combination of state and local grants can be used to deal with business development or social programs.

The guidelines for empowerment and enterprise zones change periodically, but the goals do not. In the 1960s, they were termed model cities and then urban renewal. Now, these grants are called empowerment zones. The end goal remains the improvement of the place or locality as a device to improve the opportunities for low-income residents of the place. An important example of the use of enter-prise zones today is in technology. As the New Economy companies together with the large computer corporations move to open new facilities around the country, enterprise zones are serving as a mechanism to draw the new businesses to local communities. For example, California, Massachusetts, Texas, and Florida use enterprise zones extensively to provide land for development, buildings for reha-bilitation, and job training for technology businesses willing to locate or relocate. The states actively market their enterprise zones as a way to draw local economic development to previously blighted areas. In many cases, the support of these grants has enabled the workforce to be trained and large geographical areas to be revitalized.

Tax Credits

Tax credits are provided by federal and state governments as an incentive for investment. The government does provide a grant. It uses the tax code to allow an organization to save or invest money that would be paid in taxes. Tax credits are used if investment is made for some socially useful purpose such as low-income housing, historic preservation, or research and development for specific technolo-gies or health services.

The investor will receive a multiyear credit against taxes if the development or research and development project meets certain specifications. Company XERA knows it will make a substantial profit next year, and it wants to shelter the profit from its federal tax liability. XERA may consider investing in a project with available tax credits to generate the needed tax relief.

Other Government Funding

Other government funding is provided through redevelopment laws and government agencies with funds for grants and loans alike. Urban communities in most states are subject to what is known as *redevelopment* legislation. Under redevelopment law, an area of a city can be declared a redevelopment area if it meets certain tests of blight and neglect. If the area meets the tests, it can be set aside for special treatment. In a redevelopment district, the local government (city or county) can declare private property as a public nuisance and take it for general public good at a fair price. The taking of property is called eminent domain. The taxes collected in this area are frozen at a base level at the time the area is designated as a redevelopment district.

Once redevelopment occurs and property taxes increase, the increases in property taxes or tax increment above the frozen base level are directed back to the community to pay for debt that might have been obtained for the revitalization process. The debt is typically public debt issued through tax-exempt bonds called tax allocation bonds. The sources for repayment of this debt come from the tax revenue generated above the frozen base. In certain cities and states, sales tax revenue generated within the redevelopment area may be used in the same way that property taxes are used.

Major sections of inner cities throughout the United States have used the redevelopment tools to revitalize inner cities. For example, the Baltimore Harbor project, most of downtown Los Angeles, the Van Ness Corridor and Yerba Buena Center in San Francisco, and City Center in downtown Oakland have all been built with redevelopment funds.

Redevelopment agencies use all of the available tools. They issue grants, make loans, and act as equity investors in projects. They can use their tools in the various empowerment and enterprise zones as well.

Other government programs are available for debt financing. Many of these programs are small gems, such as the Farmers' Home Administration. This little-known agency provides a number of exceptionally generous lending programs to rural and small communities. The loans cover a wide range of projects from water supplies to housing loans.

The Economic Development Administration (EDA), created in 1965, provides funds for high-profile local projects for nonmetropolitan communities. Most of the funding is allocated for infrastructure, such as roads and bridges. The infrastructure projects will create new employment.

EDA has three major types of regular funding. First, it provides grants to develop long-range economic development plans for rural areas and some major cities. These grants cover three quarters of the cost of such activity. Second, EDA provides loans for public works and development facilities. These loans carry very low rates and are usually coupled with some form of grant water purification systems, industrial parks, and roads. The grants can be used to purchase or assemble land, construct buildings, or repair existing facilities. Finally, EDA provides business loans to start or expand existing businesses located in EDA-designated redevelopment areas (distressed inner cities or rural areas).

EDA funding is fast and flexible if the community is eligible and has demonstrated capacity to use the grants and/or loans.

▬ Corporate Support

Local and national corporations have a stake in the community where they make their money. Many corporations have full-time corporate relations or corporate development officers to support a wide spectrum of local projects.

Gifts and grants are donated by many corporations for specific causes. These range from small grants to large cash donations for a project. Most corporate officers have a gift budget. In some instances, corporations donate to a special program or programmatic efforts to support local community efforts in education or community health. Corporations have taken on the ownership of specific programs or activities. Examples include but are not limited to after-school sports, community landscaping, housing repairs for the elderly, and outdoor cleanup projects.

Employee matching programs enable the corporation to follow its employees. Originally, this type of contribution was limited to higher education, where a corporation matched dollar for dollar up to a set amount for every dollar the employee donated to the university. The practice has been extended to donations for the arts, the environment, and community projects that are tax exempt. In many cases, the corporation will open this form of gifting to any community group that wins significant community and worker support.

The United Way is the most widely known national source of funding for community projects. United Way is the largest "employee-giving" program. Other Internet companies have been penetrating the field as online charitable giving has become more popular.

Employee giving programs have major impacts on local, regional, and worldwide nonprofit organizations. The programs become a continual funding source for nonprofit groups. In the case of United Way, once the nonprofit organization participates in United Way funding, it becomes a significant recipient for corporate philanthropy.

In-kind gifts and support are one of the best sources of corporate support. Many companies give items to charitable groups from excess inventory. For example, IBM, Dell, Compaq, and Apple give computers to schools and community centers. In part, such gifts are good public relations programs. However, these organizations and others similar to them are sources for funding inner-city technology centers for community groups.

Program-related investments are what companies do best. If a company can make an investment in a community housing project that pays the same yield as a Treasury Bond, it may undertake the investment. Under the terms of a program-related investment, a company invests a grant or a loan from the corporate foundation. The investment may take the form of a long-term deferred loan, stock purchase, or partnership agreement in which the company expects a return on its investment. In almost all cases, the return anticipated is somewhat lower than usual market rates of return.

Program-related investments are a growing area of corporate giving. Businesses like this type of investment because it is business and not simply philanthropic in orientation. That is, the recipient has to treat the funds as a repayable obligation. As a result, community groups have to be prepared to present the same kind of balance sheets that companies see in their alternative investments.

■ Foundation Grants

There are literally thousands of large charitable organizations in the United States. Each of these charitable organizations is called a *foundation*. Foundations exist as a result of rulings from the U.S. Department of the Treasury's Internal Revenue Service that allow organizations to accumulate tax-free income as long as their intentions are to give money away.

The foundation world is a very large growing segment of the national charitable scene. Many foundations are small, with a few million dollars as a capital base. Foundations typically spend only the interest and dividend income, keeping their principal intact so that the interest and dividend income will keep growing. Corporate foundations, such as Ford and Rockefeller, were initially founded with company stock but are now very large independent entities that have only indirect relationships with their original donors.

The large corporate foundations dispense billions of dollars annually for community and charitable giving. In some cases, the foundation restricts its giving to certain types of activities. For example, the Casey Foundation of Baltimore restricts its giving to children and family projects. In other cases, foundations restrict the geographic area for giving. The Northwest Area Foundation provides funding in the northwestern states formerly served by the railroads that traversed that area of the country. In other cases, foundations restrict the types of institutions to whom they will give. Some foundations will donate to private black colleges and universities, others will donate to girls' schools, and still others will donate to religious institutions with specific denominations. The key to raising money from foundations is to target their principal area of giving. In other words, seek money from foundations that will be receptive to your program.

The basic steps in formulating foundation requests are the following:

1. *Know what you need, and be clear about it.* If you need money for a program for unwed teen parents in Los Angeles, you start by looking for foundations in Los Angeles interested in youth services and unwed mothers.
2. *Know what the foundation has done—its track record.* Foundations publish annual reports with a listing of foundation recipients. It is best to obtain the latest report before making a funding request.
3. *Know the appropriate person to approach in the foundation.* Most foundations have project or program officers. Each of these officers has an area of exper-

tise in which he or she looks out for the foundation's best interest. If you go to the wrong person or send in a blind letter, you are likely to get a polite "no."

4. *Know the foundation rules and procedures.* Almost all foundations have a formalized set of regulations and a format for proposals. All foundations have an agenda with rules and procedures. Understand the rules and procedures before submitting a funding request. Additional information may be obtained from recipients of similar foundation grants. Annual reports will list former recipients.

5. *Know the directors of the foundation.* In small foundations, the directors are usually local people. Determine the critical issues that will appeal to the board of directors and staff.

It is relatively easy to research foundations. If you have access to the Internet, you can locate most foundations on the Council of Philanthropy Web page, www.philanthropy.com, or the Foundation Center at www.foundationcenter.org. The Web pages will provide information on foundations and directories, as well as specific data on names, addresses, contacts, and so forth. From there, you can search each foundation's individual Web site to access more in-depth information. Guides to foundations and certain types of funding sources such as community development fund-raising are located in most public libraries as well.

Illustration: Funding a Community Center and Affordable Housing Units

In a large-scale local economic development project, funding typically comes from multiple sources. In this illustration, we were requested to assemble the financing for a mixed-use development. The principal players were a private developer and a nonprofit community alliance dedicated to improving the quality of life for its residents. A public-private partnership was deemed the best corporate structure. The partnership needed to raise approximately $20 million to $50 million in grants to offset the costs of development, which totaled between $65 million and $75 million. The development consisted of a two-story retail mercado, a community center, and 450 residential apartments for low- and moderate-income families. The types of grants that could be secured are outlined in Table 6.1; types of very low-interest loans that could be secured are outlined in Table 6.2. The impact of grant funding on the financial structure of the project is outlined in Table 6.3. Without the grant funding, the project could not have been built because the rents for the uses could not provide sufficient revenues to pay conventional loans.

Table 6.1 Sources of Grants, With Areas of Philanthropy and Possible Uses for Funds

Source	Areas of Philanthropy	Possible Uses
The Sprint Foundation	Provides direct grants to support major visual and cultural activities, performing arts organizations, museums, and other cultural organizations and activities. There is a focus on effective outreach programs that broaden the cultural experience of the general public and bring cultural opportunity to underserved groups.	Amphitheater/community cultural center construction
The Prudential Foundation	Provides funds for urban and community development projects that promote public, private, and/or nonprofit sector cooperative efforts to revitalize urban neighborhoods, spur economic development, increase community stability, or solve local problems; assist in bringing disadvantaged people into the economic mainstream, primarily through the development of affordable housing, employment, and viable job-training opportunities; and promote community-based programs that foster ethnic and racial tolerance and cooperation among residents.	Affordable housing, community center
IBM Workforce Development	Funds innovative uses of technology to address key issues in Technology Grant Program job training and adult education. The organizations selected provide job training and assistance for unemployed/displaced workers, new immigrants, disadvantaged youth, and physically challenged individuals.	Community center
Department of Housing and Urban Development (HUD): Safe Neighborhood	Grants are available competitively to managers of public housing or privately owned multifamily housing for low-income people to pay for additional local police presence and other security measures to reduce crime in and around the housing.	Security for project area's residential park
Pacific Bell Foundation	Provides grants for education components of projects.	Community center's education component (amount: $10,000 to $1.5 million)

Table 6.2 Sources of Very Low-Interest Loans for Land Acquisition, Predevelopment Costs, and Gap Financing for Construction, With Areas of Philanthropy and Possible Uses for Loans

Source	Areas of Philanthropy	Possible Uses for Loans
Commercial Banks: Bank of America, Wells Fargo Bank, etc.	Provide loans to owners, nonprofit developers, and for-profit developers to rehabilitate and/or develop new housing projects and other community facilities in underserved communities.	Construction loans for residential units and community center
City Housing Department	Provides loans to owners, nonprofit developers, and for-profit developers to rehabilitate and/or develop new housing projects using federal funds. The types of loans offered are acquisition, predevelopment, construction, and permanent financing (gap) and bridge assistance.	Low- and moderate-income housing
Section 221(d)(4) Mortgage Insurance	FHA (part of HUD) insures loans originated by private, HUD-approved lenders. Prospective project sponsors/mortgagors are responsible for finding a HUD-approved lender to make a loan. Section 221(d) is not a direct-loan program.	Affordable housing permanent financing
Low-Income Housing Tax Credit Program	Provides private developers with federal and state tax credit if a certain percentage of the units in the projects are rented to low-income occupants.	Low- and moderate-income housing equity investment

Table 6.3 Impact of Grants: Comparison Matrix (in $)

	Best Case	Most Likely Case	Worst Case
Rents (per Sq. Ft.)			
Retail space	2.25	2.00	7.85
Community center	1.00	0.85	0.75
Moderate-income apartments	1.35	1.25	1.15
Low-income apartments	1.10	0.90	0.85
Monthly vending revenues	30,000	22,000	16,000
Monthly parking revenues	43,250	43,250	41,250
Monthly vacancy	3%	4%	5%
Debt/Equity Without Grants			
Total Cost of Project	72,545,000	65,780,000	70,005,000
Equity in Project Without Grants	17,932,950	15,396,500	15,605,450
Loan Amount Without Grants	54,612,050	50,383,500	54,399,550
Grants			
Land purchase and development costs	11,000,000	5,500,000	2,750,000
(All of the cost)	(1/2 of the cost)	(1/4 of the cost)	
Community center	3,775,000	3,775,000	2,831,250
(All of construction cost)	(All of construction cost)	(3/4 of construction cost)	
Low-income family apartments	17,437,500	15,500,000	11,625,000
(2/3 of construction cost)	(3/4 of construction cost)	(1/2 of construction cost)	
Parking	5,197,500	4,320,000	3,465,000
(2/3 of construction cost)	(1/2 of construction cost)	(1/2 of construction cost)	
Moderate-income family apartments	8,500,000	4,250,000	—
(1/2 of construction cost)	(1/4 of construction cost)		
Total Value of Grants	45,910,000	33,345,000	20,671,250
Debt/Equity With Grants			
Equity in Project	5,327,000	9,096,250	16,444,583
Loan Amount	21,308,000	27,288,750	32,889,167
Equity in Project	1/5 of project cost	1/4 of project cost	1/3 of project cost

Requests for Funds

Requests for funds are formalized in two distinct formats: the strategic business plan and the proposal. The strategic business plan, as outlined in Chapter 2, is primarily used for equity and debt funding. The proposal is used for seed money and grant funding from public agencies and foundations. The seed money provided by agencies and foundations will often provide sufficient resources to prepare a professional business plan. Once the business plan is complete, the organization should forward a copy of it to the foundation or local agency that supplied the initial funding for the start-up enterprise.

The business plan components are the subject of this book. Outlined below are the necessary steps for preparing a proposal to solicit foundation seed money or initial funding for an economic development start-up enterprise. Proposals have seven components. Each is important and should be fully developed and documented.

▬ Summary

The summary is a concise presentation of the project and states its purpose. The summary, no more than two pages, explains to the foundation or government agency who you are and why you are credible to perform the tasks outlined in the document. Identifying the key professionals and the board of directors is essential so that the foundation can evaluate your stature and experience. The proposal should be written on your letterhead.

Establish with some supporting data the reason for the grant request, issues, problems that are being addressed by the project, the objectives to be achieved, and the means used to accomplish these goals. Finally, the summary must identify the total cost of the project, the funding request, and the use of the money should the foundation provide a grant or a loan.

▬ Problem Statement

The problem statement provides the detailed data on the project based on some scientific reasoning. The data supporting the statement of need must come from credible sources. If the proposal states that child abuse is an important problem because of national reports on the issue, the proposal writer will lose points. The proposal author must explain the degree of child abuse in the local area and support the findings with quantitative evidence. Local data and statistics should be augmented with local newspaper articles or quotations from local authorities to provide added credibility.

The problem statement must be clear so that the reader learns something from the statement that is interesting and not merely demonstrative of the gravity of the issues. For example, if you say that your local area has the worst child abuse problem in the nation, the point may be important but is not convincing. On the other hand, saying that you have traced the community's child abuse problems to the recent plant closings gives the reader pause.

▬ Objectives

Objectives establish the benefits of the program in measurable terms. The objectives should be stated in terms that indicate how much will happen and when. A proposal must clearly state the goal of the organization. For example, the goal of placing 400 disadvantaged youth from Youthville in college over the next 3 years is concise and can be measured. In simple terms, the proposal has to explain what methods will be used to achieve the goals that are formulated by the new enterprise. If the proposal states that an intensive language and math program developed in China will be used to prepare students for college, the methods are clearly stated.

▬ Methods or Approach

This section describes the activities used in the program to meet the stated objectives. Methods statements need to be specific enough to convey the idea without going into all of the details. The details will be covered in the business plan. In the case of the proposal to use a language and math program developed in China, the methods might include an evening session on the abacus and Chinese learning games to teach symbolic logic. If the section refers to remarkable scores on conventional exams of American students who have attended school in China for a year or more, this may attract the attention of the grant officers. If there has been an experiment in the local schools supervised by the local university, then the methodology may seem plausible.

▬ Evaluation

Document your evaluation process and how you will measure goals and the completion of these goals. In some cases, external reviews are required. In other cases, taking data from the program participants will provide the required evidence. In any case, the methods and the evaluation systems need to coincide so that adequate reports can be filed with the foundation(s) supporting your project.

▬ Other Funding Sources

Provide the foundation with information on other sources of funds. Foundations generally prefer to share risks with other lenders and grant makers so that they are not assuming all the risk. Foundations need to know how the project will be funded after the initial start-up. Providing the financial plan will ensure that the foundation has comprehensive information.

▬ Budget

This section must present a global summary: the bottom line in terms of revenues, fixed costs, and variable costs. Budgets for development projects require detailed documentation for construction costs and construction loans, as explained in Chapter 5. Foundations, government agencies, and local philanthropic groups require detailed multiyear budgets, sources and uses of funds, and any backup data for large planned expenditures.

Experienced foundation officers know if the project budget is under- or over-inflated. These entities provide funding for many projects and will compare your budget to those of similar projects. Government funding is even more sensitive in this regard because your project will be tied to the budget you submit. If you over-estimate, you run the risk of being disqualified. If you underestimate, you may not go back and request additional funding because your costs came in higher than budgeted.

Budgets for any business plan are very similar to the start-up and operating pro formas detailed in Chapter 5.

▬ Letters of Support or Collaboration

Letters of support or collaboration from government, community groups, or significant leaders enhance the proposal. These letters need to be specific and tell how much the writer knows about the organization and the project. Reasons for support must be explicit.

Conclusion

The purpose for accessing money from multiple sources is to ensure that your project is funded. Typically, one source of funds will not provide sufficient investment for any project. Layering the funds will secure what is needed.

In summary, the first level of funding is seed money secured from foundations, grants, the entrepreneur, and the local community. The initial equity funding is the second level secured from the initial equity investors. These investors assist the organization with funds for research and development, start-up operations, and the preparation of the business plan. The third level of funding is from a broader base of equity investors or venture capitalists. These groups provide a sufficient amount of equity (20% to 30%) to ensure that debt funding is obtainable. The final step in accessing the funding is to obtain debt financing from institutional lenders and government agencies. These creditors will require a detailed business plan for the program, project, or service and the requisite financial statements to ensure that overall risk is minimized.

Case Study 6: Brava Valley Project Funding

The Brava Valley Community Corporation (BVCC) has requested that you seek foundation and grant money for the child care and elder care projects. They will require funding for the community center as well. The BVCC has also requested that you identify sources of funding for technology businesses because they will use part of the property for a technology business center. Using the results of your cases in Chapters 4 and 5, provide the following:

1. *Access funding sources.* Search the Internet to determine the funding sources for the above projects. Use the Web sites identified earlier in this chapter. List the sources of funding and how each will assist you with your project.

2. *Prepare the budget.* Prepare a preliminary 1-year budget for your project, using first Table 6.4. After you have completed Table 6.3, prepare a detailed multiyear budget, using Tables 5.5, 5.6, and 5.7 as examples. Show the total cost of your child care or elder care project for Brava Valley and then for the new technology center. Expand your budget to include 5 years of operations and the proposed grants or foundation loans you will receive.

3. *Write a proposal letter.* Prepare a proposal letter to a foundation requesting funding for your project. Include a summary of the project, a description of the organization developing the project, and why the project is important. Be sure to identify the amount of money you are requesting and how the funds will be used.

Table 6.4 First-Year Budget Components and Sources of Funds

	Total Required	Source A	Source B	Source C	Source D
Payroll and taxes					
Contract services					
Rent and utilities					
Equipment and leases					
Supplies					
Insurance					
Travel					
Legal and accounting					
Miscellaneous					

Grand Totals:

Putting the
Plan Together

The strategic business plan is the most critical document for local economic development. The plan gives the equity partners, the lenders, and local community investors a comprehensive approach to the desired program. With a detailed plan in place, all interested parties can examine how the product or service will be developed, how the organization will be structured, who will be funding the program, how the organization will market and implement the plan, and what benefits and risks exist for all the stakeholders. Stakeholders are critical to the entire process because with their time, energy, and capital funding come expectations that must be met. To that end, the strategic business plan enables the community entrepreneurs and business participants to come together with the local community public and private leaders to forge a local economic development plan that works for all groups.

The steps to put the strategic business plan together are as follows:

1. *Document the strategic intent and the goals.* Define the overall program, and outline the mission of the project. Specificity is essential because the strategic intent will be the guide throughout the entire planning process. Address the goals that you want to achieve, how they will be achieved, who will take responsibility, and what groups will be accountable.

2. *Prepare a summary of the organization plan and the key professionals.* What is the structure of your organization? Who are the partners or executives? How will the organization operate? Who is accountable for the multiple decisions that must be made as the organization is formed and becomes operational? What do you need in terms of people, time, money, and equipment?

3. *Provide details on the market.* Who are your customers, and where are they located? Who is your competition? How will the competition affect your product or service? How will you market your product? What is your sales forecast for 1 year and for 5 years? Understanding the market will enable you to target the appropriate resources to be competitive and successful.

4. *Document the financial plan.* Include start-up costs, 5-year pro forma, budget, and sources and uses of funds. Who are your major investors and lenders? What are your sources of grant funding? How will you project cash flow? What are the yields to the investors? How much money do you need? Where will you get additional capital if you have cost overruns or burn cash too quickly?

5. *Describe the implementation plan, and identify stakeholder benefits and risks.* Who wins and who loses? How will your project or service be implemented? What is your time line, and what are your critical dates? Who will be responsible to ensure that the organization meets its critical dates? Examples of the stakeholder benefits and risks are provided in Tables 7.1, 7.2, and 7.3.

Illustrations of executive summaries are provided below. They cover different ventures with both simple and complex organization structures.

In the Triangle Project, we have charted the benefits and risks from a technology center. Clearly, all the stakeholders—ABC Technology executives, the conventional lender, the venture capital groups, the international businesses and trade groups, the community residents, and the city—win in the process. The benefits include increased jobs and higher salaries for the labor force, tax revenues for the city, and profits for the technology company, the lender, and the venture capitalists.

In ACC/TBC Public Services, we have illustrated the benefits and risks associated with the proposed service development discussed in Chapter 6 and documented in Table 7.2. This project will have substantial benefits for the community but will involve risks associated with the potential for insufficient cash flow. The benefits are very great for the residents and the city, however. The downside exposure is $10 million provided by the foundations.

Once all the components of the plan are well documented, it is essential to prepare an executive summary covering the key points in the plan. The executive summary is the first section of the strategic business plan and provides the reader with an overview of the key findings and strategies. Typically, the executive summary can be three to four pages long or longer, outlining the important organizational framework, the market summary, key financial results, and the benefits and risks to the stakeholders. The executive summary will also state the amount of money being requested by the new enterprise.

(text continues on page 156)

Illustration 1: Executive Summary
for the Triangle Project

KBJ's Triangle will foster growth opportunities in the community it serves. KBJ is a nonprofit development corporation committed to working with cities to build projects that benefit the community. Building the Triangle will create a brighter future for the residents and businesses it serves. The Triangle will be a three-story development located on the corner of Scenic Boulevard and San Tomas Boulevard, west of the downtown. The development will combine education, technology, and business components, forming a triad of uses to create a more unique urban center. Coupled with the city transit program and a new transport line, the Triangle will turn into a destination point for incoming technology workers from outlying areas. The combination of all these uses in an architecturally significant complex will create an urban economic catalyst that will increase the quality of business and economic growth.

The Triangle Center will assist workers to prepare for the future, knowing that population, economic, and industry changes are taking place. As the local neighborhoods continue to grow at a faster rate than the rest of the city, more young people are entering the community by way of job creation.

A detailed analysis of the community needs and the city's goals has prompted us to come up with a specific product, anchored by a high-technology education and startup business/ABC Technology. The implementation strategy is based on striking a series of productive partnerships between various community and regional stakeholders. The land acquisition would be completed through a partnership with the Redevelopment Agency granting a long-term lease agreement. The nonprofit sections would be funded by numerous grants from government and private stakeholders. The equity for the for-profit part would be gathered through the State of Nevada Employees Pension Fund and Tax Increment Redevelopment Bonds. The equity for the technology business will come from venture capitalists. In addition, a Section 100 loan and a loan from the Community Development Bank would be taken to fulfill the remaining financial needs. A detailed financial analysis considering both high- and moderate-growth scenarios shows positive internal rates of return at the end of 5 years (in a most likely case). This makes the project highly feasible on a long-term basis. It is extremely encouraging to see that even the modest assumptions show a long-term viability of the project. With its range of beneficiaries, the project could set an example to other neighborhoods, communities, and cities to come up with similar innovative solutions based on effective partnerships to even out economic and inequities and increase opportunities (see Table 7.1).

Table 7.1 Benefits and Risks From Triangle Technology Start-Up

Stakeholders	ABC Technology Execs.	Conventional Lender	Venture Capital Groups	International Businesses and Trade Groups	Community Residents	City
Investment	10% equity investment of total product and start-up costs	Loan for 55% of total program costs	35% of equity required	Marketing and public relations	Labor force	Job-training funds
Return (%)	50% cash-on-cash return; value of stock and equity to be determined after vesting—4 years out	11% interest	Stock and equity—value to be determined—est. ranges from 100% to 225%	Standard fees only		
Cash flow	$21M after-tax cash flow, with $1.17M profit	$695,000 on 24-month loan	$35 million after IPO			
Benefits	Profit; long-term economic health of corporation	Guaranteed loan payments provided by "angel investors"	Money	Significant opportunities for international trade	Increased benefits from higher salaries	Tax revenues on new properties; multiplier effects from workforce
Risks	Time-to-market longer; competition; cash flow; return on investment	None—first call on funds	Potential loss of investment	Potential loss of fees	Increased traffic, noise and parking issues	Potential for concept not being realized

Stakeholders

Government Agencies

Government agency investments would be in the form of land acquisition, re-development loans through tax-exempt bonds, and $750,000 in grants for the job training. There would be a cash flow of $115,000 annually from the land lease. IRR is not calculated.

Benefits would be an increased sales tax base, increased quality of life for the community, and a catalyst for development for downtown improvement. The primary risk would be potential default on the bond payments.

Nonprofit Developer

The developer's investment would be in the form of management expertise: taking responsibility for raising equity and obtaining grants from foundations. Cash flow is likely to be $304,000 annually.

Benefits would be a fulfilled vision, community contribution to improved business expansion, and positive rates of return, permitting reinvestment into the nonprofit organization. Risks would be inability to obtain grant funding, poor rents, very slow absorption of space, and business failure. Other risks are assumed by ABC Technology and include longer time to market, lower cash flow, and lower return on investment.

Equity Provider

The equity provider's investment would be $1.4 million.

Benefits would be increased availability of trained labor and the fulfillment of technology business goals. The major risk would be business failure of the project, leading to goal failure. While the return on the development deal may be low, the returns on the technology business are very high—50% cash or cash return.

Illustration 2: Executive Summary for the ACC/TBC Public Services Program

ACC has been among the communities hardest hit by economic downturns. Its labor pool is primarily unskilled, and its downtown has marginal commercial uses, empty lots, boarded-up storefronts, and an aging or deteriorated housing stock. The decline in the physical environment is mirrored by social indicators of a community in distress. The local populace is characterized by low educational attainment, poor English-speaking ability, and a high rate of poverty.

Our study proposes the formation of an organization that will deliver four major services to the community. These services include child care, medical care, an art and technical training school, and a community park. The organization will be structured as a public/private joint venture partnership between ACC, the non-profit economic development corporation, and TBC, a for-profit service provider. The organization will be dedicated to providing essential social services and job training for the community. (See Table 7.2.)

Stakeholders

The stakeholders in the Service Plan include the community members, the city, the public and private lending institutions, and the for-profit group, TBC, which will deliver the services.

Investment

All of the proposed services are heavily dependent upon grants and loans from government agencies and foundations.

Returns

If all four services are completed, the internal rate of return will be 7.8%. Rates of return vary from project to project, with the park having only expenses and no revenues other than community park fees that are part of the residential tax bill. Optimally, the services will be evaluated by foundations and grant makers on the basis of need and benefits to the larger community rather than financial performance.

Time Line

Time lines are crucial in any major project so that all relevant agencies are aware of their responsibilities and duties at any given point throughout the development process. Common critical points are found throughout each project. These critical measures include the finalization of planning for needed contract services and the finding of a suitable building for renovation. City assistance is essential for this first step. Other critical measures include the securing of grants, private debt, and public debt.

Benefits and Risks

The benefits that accrue to the stakeholders reflect the goals of the proposed organization formed by ACC and TBC. These include the elimination of blight; the provision of sufficient job training, child care services, and medical services; and the development of a park for families and children.

The primary risk is that the proper organizational structure will not be established to support the development of the social services. In such an event, additional costly outside contract services will need to be retained to oversee the project (Figure 7.1).

Table 7.2 Benefits and Risks From ACC/TBC Public Services

Stakeholders	ACC/TBC	Multiple Lenders	Foundations	Merchants	Residents	City
Investment	N/A	$2 million	$10 million			
5th-year return (%)	7.8%	9.9% combined interest	No financial return			
5th-year cash flow	$4.9 million					
Benefits	Profit; long-term social and economic health of community	Community investment plus return of capital plus interest	Community investment	New retail facilities; opportunity for increased profits	New and affordable housing	Increased tax base
Risks	Failure to provide sufficient cash flow for projects	Failure to obtain return of capital	Potential loss of investment	Business failure	Project defaults	Potential for concept not being realized

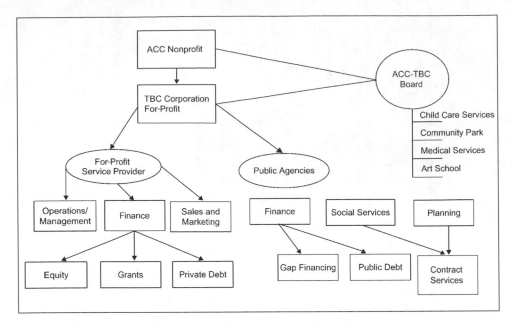

Figure 7.1. Organization Structure and Sources of Funds

Illustration 3: Executive Summary for the Riverfront Retail Development

This document has been prepared by MC Corporation and outlines a development proposal for a 365,000-square-foot theme retail development in the city of Rivers. The project will be named The Riverfront at Corner Creek. Specifically, a department store, a discount clothier, and a nine-screen theater complex will anchor this development. Additional tenants include casual dining restaurants, a quick-service food court, and in-line retail shops.

Market demographics support the development. From our analysis of key demographics within a 3-, 5-, and 7-mile radius of the project site, the surrounding market area consists of a high-density, diverse population expected to grow at moderate rates over the next 4 years (between 3% and 4.5%). Growth in median income for market area households is expected to remain stable, between $31,000 and $33,000. We believe that this projected increase in population, along with a relatively stable median household income, will increase the demand for retail goods and services in the area over the next 10 years or longer.

Consumption patterns of market area residents indicate the need to serve multiple retail categories such as the following:

- Auto and home supplies
- Gasoline/service stations

- Hardware, lumber, and garden supplies
- Food stores
- Department stores
- Drug and proprietary stores

Given that department stores can capture retail dollars in multiple categories, we believe that the consumption patterns identified above support the development proposal to include a major department store as an anchor tenant.

A retail gap analysis of the supply of retail and service establishments indicate that niche opportunities exist in the following areas:

- Miscellaneous retail
- Apparel and accessory
- Eating and drinking
- Legal services
- Motion picture and amusement
- Health services
- Business services

MC Corporation believes that our development proposal will prevent further leakage of retail sales to outlying communities and enable our city to service the needs of the residents, specifically by providing a discount clothier, movie theaters, and retail that targets these niche opportunities.

The development proposal will result in positive after-tax cash flows between $1.8 million and $2.3 million. Additional financial analysis indicates that the project is financially feasible, with an internal rate of return of 21% at the end of Year 5. Though a leasehold option is feasible, it is not a recommended strategy because the internal rate of return will turn positive only after a 9-year period.

To meet the goals and guidelines of the city, we recommend the creation of a public-private partnership between the developer, the city, and the Community Redevelopment Agency. The basis of this recommendation is to provide access to additional funds for development, as well as to begin the process of fostering a more positive image for the city and the redevelopment agency. A positive image for the city will attract additional development into the area in the future.

Although the developer and the Community Redevelopment Agency will bear the most risk as public-private partners in the development of the project, there are numerous benefits that can result. These are listed in Table 7.3.

As a marketing and real estate development company specializing in the revitalization of urban areas, MC Corporation is grateful for the opportunity to submit this business plan for funding consideration.

Table 7.3 Benefits and Risks From Riverfront Retail Development

Stakeholders	Development Corp.	Investor Partner/ Lender	Redevelopment Agency	City	Retail Tenants	Residents
Investment	$16.3 million	Loan of 65% of total	Land write-down of $16.5 million tax increment financing	Staff and administration	Retail Association fee of $0.35 per sq. ft.	
5th-year return (%)	21%	6% blended interest rate				
Benefits	Return on investment	Loan repayment plus interest	Improved land; improved image; additional tax increment; potential for additional development within redevelopment area	Sales tax revenues; positive image; regional draw to retail development; significant use of underutilized property	Profit potential; collective marketing	Retailers to provide needed goods and services; positive image of city; potential for increased property values
Risks	Failure to provide sufficient cash flow; business failure; inability to repay loans	Loan default	Developer pulls out of project; default on loan	Developer pulls out of project; default on loan	Changes in consumer preferences; competition	Traffic, noise

Illustration 4: Achieving an Alliance for the Community

The Community Alliance is a nonprofit, community-based organization dedicated to improving the quality of life for the residents in Middle Town, a community 30 miles from Brava Valley. Middle Town is a well-established community, densely populated, and with a wide range of community demographics. Middle Town has a small-to-moderate group of affluent residents, a large group of middle-income family households, and a moderate-sized group of lower-income households. The Alliance seeks to achieve the goal of improving the quality of life for *all* residents through active community participation and equal representation of all community groups. The structure of the organization includes a board of directors, an executive management team, and a local community committee. The Alliance, together with its developers, will build a catalyst project in the Middle Town Community Planning Area (MTCPA). This mixed-use project will cost $23 million.

The purpose of this strategic business plan is to seek funding from the city in the form of land write-downs. We are also seeking funds from public purpose foundations to underwrite some of the cost of senior services and job training for the citizens of Middle Town. Our total request for funds for services and job training is $3 million. We are requesting $3 million for land cost reduction. The Community Alliance will pay the balance of the funds, or $2 million, as equity.

Vision Statement

The Middle Town Community Alliance will serve as the catalyst for local economic development, provide low- to moderate-income housing, and create a community dialogue and gathering place for the residents of Middle Town.

Goals and Objectives

The Community Alliance will endeavor to achieve the following:

- Create wealth within the local community.
- Reinvest the profits of any development back into the community.
- Provide economic development and employment within the area.
- Provide needed retail and other services in addition to low- and moderate-income apartment units for the diverse population.
- Facilitate dialogue and cooperation between the varied racial, ethnic, and religious groups.
- Provide community services and facilities.
- Provide open space.
- Create a unique architectural landmark and land use within the community.

Site Location and Context

Old, intensively developed commercial districts surrounded by densely populated residential neighborhoods characterize Middle Town. Historically, commercial development has occurred along the major transportation corridors, whereas residential development has been located on the blocks behind the main thoroughfares.

The Community Alliance site is located approximately 2 miles from Middle Town's downtown, 15 miles from First City, and 30 miles from Brava Valley. The site covers an entire city block and is approximately 195,000 square feet in area. At present, a five-story medium-sized office building and an apparel store occupy the northeast corner of the site. The balance of the site is vacant.

Significant land uses adjacent to the site include a midsized commercial office building, a historic landmark, Our Lady of Mercy Catholic Church, and a cultural center and museum that provide historic information, maps, and art covering the American Indian population and the Mexican American population. The immediate area to the south of the site has fairly dense residential land uses with multifamily apartments and single-family homes. The southern boundary of the site has a mix of residential and commercial uses, including older mixed-use buildings. The retail outlets, including art galleries and restaurants, display the distinct ethnic influences of the region.

The area has access to limited public transportation. Most of the community travels by foot or by car. Bus transportation is available.

The Community Plan for Middle Town indicates that the site is zoned for commercial use, residential use, and parking. The allowable floor area ratio (FAR) is 6.0. The site also falls within the boundaries of the Community Redevelopment Agency's Project Area. The city of Middle Town envisions an open-air market catering to local ethnic populations. Eventually, the street may be closed, and the market may be expanded to other surrounding streets.

Development Scenario

The preliminary market analysis indicates that the area is deficient in residential and community services. Visualized as a community marketplace or *mercado* by the city, the development scenario for the site seeks to complement this vision with additional uses needed by the community in the near future.

The scenario includes a dense, mixed-use development on the site with retail, residential, and community services. The development will be called the Community Courtyard. The architectural style envisioned is southwestern/Spanish with a contemporary theme. The proposal also includes a clock tower to complement the landmark character of the local religious buildings and ethnic museum and art center. Extensive landscaping, fountains, and walkways are proposed to create a pleasant ambience with the Community Courtyard.

Community Center

A community center of approximately 65,000 square feet (four stories) is proposed in the center of the site, acting as a buffer between the residential and retail components at the ground level. To the immediate west of the site is an open green space that would complement some of the activities and uses within the community center.

Some of the facilities incorporated within the community center would include

- An Interfaith Center to act as a forum for discussions among the varied religious and ethnic groups within the community and to bring about greater understanding of different cultures and religions
- An amphitheater, which would be the heart of the development. It would be used for theatrical and musical performances, both public and private, during lunch and evening to increase revenues for the community center. It could also be used as a public square when not being used for performances.
- Multiple-use meeting or lecture rooms for language training, computer training, other job skills training, a senior services center, community meetings, and art exhibitions
- An indoor/outdoor basketball court with supporting gymnasium and locker rooms
- Administrative offices of the Community Alliance

Retail Center

A two-story retail complex, approximately 65,000 square feet in area, is proposed on the northern edge of the site. The main concept driving the project would be to create a thematic festival/promenade marketplace at the street level with apartments above. The entry to the retail complex could be through the proposed watchtower to distinguish it architecturally and increase its visibility. Some of the proposed tenants include

- Three to four ethnic restaurants (5,000 square feet each)
- Restaurant with night club or sports bar (10,000 square feet)
- Book/music/gift store catering to the majority ethnic groups in the area (25,000 square feet)
- Art gallery (5,000 square feet)
- Vendors such as a newspaper stand, a florist, and crafts (5,000 square feet)

These retail establishments are expected to attract the residents in the vicinity as well as office workers in the area. The project will also attract visitors from neighboring communities.

Residential Apartments

As part of the residential component, low- and moderate-income apartments are proposed for the development. Low-rent apartments would be to the south of the site, and moderate-rent apartments would be located above the retail component. There would be approximately 450 dwelling units in all, with a mix of one-bedroom, two-bedroom, and three-bedroom apartments to attract a cross section of renters.

A terraced building roof line with roof and terrace gardens, a landscaped court-yard with a fountain, covered parking, and ample security are some of the amenities envisioned for the residential apartments. The structures would vary from between four and six stories and would be approximately 510,000 square feet in area.

Apartment Description

Low-income units (310,000 sq. ft.)	Moderate-income units (200,000 sq. ft.)
One bedroom: 75 (650 sq. ft. each)	Studio: 50 (500 sq. ft. each)
Two bedroom: 100 (1,100 sq. ft. each)	One bedroom: 75 (700 sq. ft. each)
Three bedroom: 75 (1,300 sq. ft. each)	Two bedroom: 75 (1,200 sq. ft. each)
Total no. of units: 250	Total no. of units: 200

Parking

Due to the large number of residential units and the retail planned for the site, approximately 825 parking spaces will be required. This parking proposal assumes that the proximity to bus stops will reduce the parking requirements. Most importantly, the community center is primarily targeted to residents in the immediate vicinity of the project site; therefore, our plan anticipates more pedestrian traffic rather than vehicle traffic. Parking requirements for the community center are assumed to be low.

The parking structure will be six stories high to accommodate 725 spaces. It will be constructed on the east side of the site. Parking will be provided on the ground level, increasing capacity by an additional 100 spaces.

Parking

Parking requirements	825 spaces
Retail (3 spaces/1,000 sq. ft. rentable space)	195 spaces
Residential (1.5 cars per dwelling unit)	600 spaces
Community center	30 spaces

Accommodating retail parking in the adjacent parking structure across from the site can compensate for the deficit in parking on the site. Access and egress to the parking facilities will need to be designed, keeping in mind that Middle Town may eventually close some of the streets in the neighborhood.

Market Analysis

Site and Area Analysis

Because of its size and location, the project site is physically suitable for the development of this mixed-use project. The site can accommodate the construction of a sizable, multiuse project. It will be highly visible and accessible to pedestrian, transit, and automobile traffic. The surrounding historic buildings enhance the visibility and character of the site.

Demographic Analysis

The primary market area for Community Courtyard is determined to be the Middle Town Community Plan Area (MTCPA). However, there are slight variations as to the number and type of people that each component of the project will draw.

According to the 2000 census, the total population of the MTCPA was 271,001, and the total number of households was 106,430. Tract ABCD, in which the courtyard is specifically located, had 5,482 persons. The primary market for the retail component of the project is this census tract along with the adjacent tracts. Office workers along Middle Town's central business district are also expected to be the primary patrons of the shops and restaurants. The residential population and the employment population have very different incomes and spending patterns. This business plan attempts to reflect the varied users and their consumption patterns.

Additional data indicate the following:

- Average household income for the MTCPA is $32,245; the average household income for the adjacent areas is $24,142.
- Hispanics make up nearly 37% of the population in the MTCPA; whites (non-Hispanic), 45%; Native Americans, 10%; African Americans, 5%; and Asians, 3%.
- The annual population growth and housing growth for the planning area have been higher than the citywide average.
- Population density is three times higher in MTCPA than the citywide average.
- Approximately 14% and 86% of the housing units in the planning area are single-family and multifamily units, respectively.
- The 2000 census indicates that 82% of the residents in MTCPA are renters and only 18% own their own homes.
- Approximately 62% of the MTCPA population is foreign born, whereas the citywide average is 45%.

Demand and Supply Analysis

Retail Complex

There is little diversity in the retail and restaurant establishments in the area. Though no secondary data existed to prepare this research, we undertook a lengthy survey of 90% of the residents and workers in the MTCPA and 85% of the office workers in the central business district. Most of the office workers interviewed by the project team complained of a lack of shops and eateries in the immediate area surrounding the site. Apparently, even though there are quite a few eating places along the main central business district strip that cater to the lunchtime office crowd, the workers demand a greater number and variety of eateries. Residents of the area stated that there were insufficient quality restaurants in the area to serve their individual and family needs. The price points mentioned by the office workers ranged from $3.75 to $5.00 for lunch entrées exclusive of beverages and $6.00 to $8.00 for dinner entrées exclusive of beverages. To satisfy such demand, the proposed project will include a variety of ethnic restaurants that are lacking in the area. Restaurants offering Middle Eastern, Guatemalan, Salvadoran, and Chinese high-quality food at moderate prices will attract both office workers and nearby residents.

In terms of retail, specialty shops such as bookstores, music shops, and gift stores that complement each other can expect to have a sizable clientele. The retail component of the development aims to tap into the growing buying power of the underserved residential neighborhoods around the site.

Residential Apartments

There is demand for low- and moderate-income rental housing in the area. New units are needed to meet the demand for housing by new or recent immigrants, existing area residents desiring to move into newer and/or larger units, and residents from other areas. Both the income and housing characteristics of the area justify the development of the housing units that we propose. Our research indicates that 70% of the area households have very low to moderate income. The problems of overcrowding and the deterioration of existing housing also suggest that new units are likely to be in demand by households in the area who want and can afford to move.

Household Income

Household Income Categories	Project Area	City of Middle Town
Very low (2% of county median)	28.8%	22.6%
Low (50% to 80% of county median)	20.2%	15.7%
Moderate (80% to 120% of county median)	21.0%	25.9%
Total	70.0%	64.2%

The vacancy rate for apartments in the area was 5.0% in 2000. This is a positive sign, as a vacancy rate of 5% to 6% generally indicates a more fluid and healthy housing market. Some of the buildings in the area have no vacancy at all. No new apartment constructions were observed or are planned for development over the next 5 years.

Community Center

During the primary field observation and interviews with the local residents and employees, we found that there was substantial interest in seeing a community center developed on the site. The closest community center is 5 miles away and on the other side of town.

A new community center developed on the site would complement the existing centers by providing services and facilities currently unavailable or lacking in the area, including special youth-oriented facilities, community meeting rooms, job and computer training services, and facilities for the elderly, who currently lack any facilities within walking distance.

Office Space

At the end of 2000, Middle Town had an office vacancy rate of 7%, down from 15% at the end of 1999. Middle Town has experienced high growth in computer software and media businesses that has resulted in a strong demand for local office space. The 2000 asking rents were $1.30 and $0.75 for Class A and Class B spaces. Because of construction costs, we do not believe that these rents will cover our costs. Therefore, office space will not be a component of the project.

Management

The Community Courtyard will be operated by a three-member executive management team under the supervision of the board of directors. The executive management team will supervise the construction and the day-to-day operations of the development. The board of directors will direct the management team's responsibilities within the development, and a community advisory committee will provide input to the board of directors through the executive management team.

Executive Management Team

The salaried executive management team will consist of three members. The management team will supervise construction and maintenance, manage the property, collect rents, manage and operate the community center/amphitheater, administer long-range planning, coordinate financial operations, solicit input from stakeholders (community members, local businesses, local religious institutions, local government, etc.), and coordinate and chair the community advisory committee.

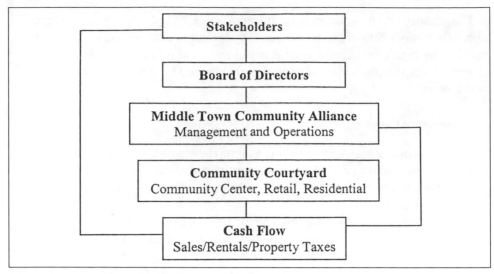

Figure 7.2. Project Structure

The executive director will be in charge of coordinating overall operations, planning, and financing/fund-raising. He or she will represent the team at the board of directors meetings as well as chair the community advisory committee. One of the other team members will be in charge of property management and operations of the entire development. The remaining team member will be in charge of the operations of the community center/amphitheater.

Board of Directors

The board of directors will consist of up to seven members. They will be appointed/elected by their respective constituents. The board will represent the city of Middle Town; equity partners; grant providers; local businesses; local community members, including religious leaders; the executive management team; and the commercial lender of low-interest loans. Another community member will replace the lender once the loan is paid off.

The original board will be appointed by the stakeholders. After an initial period of 1 year, the board members will be either reappointed or replaced by appointment or election by their respective groups. After that, each board member will serve a 2-year term not to exceed two consecutive terms.

Community Advisory Committee

The community advisory committee will be composed of local stakeholders. A list of acceptable members will be cleared through the board of directors by the executive director of the management team. These members will then vote on a 15-member committee to represent community interests, provide input to the

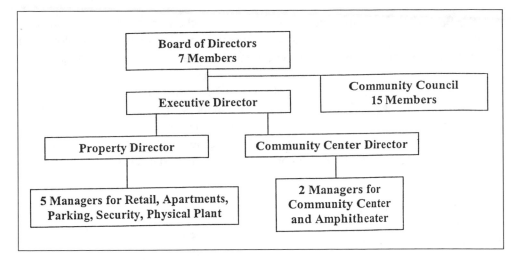

Figure 7.3. Project Management

management team and the board of directors, and appoint the community and local business members of the board of directors.

Financial Operations

We have conducted a series of cash flow analyses using various assumptions, debt instruments, grants, and incentives from the city of Middle Town. [We refer the reader back to Chapter 6, Table 6.3, for an example of the impact of grants and low-interest loans on the project.] The grants for Community Courtyard have come from a variety of sources, including the U.S. Department of Housing and Urban Development (HUD), the State Community Foundation, XYZ Technology Corporation, and many others. The public and private agencies and businesses will contribute $3 million for the reduction of construction costs. The city of Middle Town will issue tax increment bonds for $15 million to cover the long-term debt on the project.

Cash flow generated by the retail center will provide Middle Town with increased sales tax revenue. Increased property taxes will provide the redevelopment agency with funds to service the bond debt through tax increment financing. All parties will receive nonmonetary benefits through community revitalization.

The start-up costs in the first year are estimated to be $2,050,000. These costs create an operating deficit in the first year. This deficit will be covered by a commercial loan at a 10% interest rate, to be paid off in the 2nd year of operation.

Financial Scenarios

We prepared three financial scenarios covering the most likely case, the best case, and the worst case to test the financial viability of the Community Courtyard. On the basis of engineering figures, we were provided quotations for total construc-tion costs that equaled $28 million. Major technology companies wrote down our costs to $23 million so that they could support the growth of technology in the region.

We have provided a summary of the key financial statistics below. We believe that our most likely scenario is conservative in its assumptions, though optimistic about the Community Courtyard's attraction of potential tenants. It assumes cur-rent comparable market rents for all of the components. Vacancy rates also are reflective of the present market conditions. The blended interest rate is optimistic at 7%.

Investment Returns

Statistics	Year 3	Year 5	Year 6
Net operating income	$3.100 million	$3.883 million	$5.500 million
Debt coverage ratio	1.92	2.40	3.40
Cash flow before taxes (after debt)	$974,415	$1,126,158	$2,052,096
Net present value	($2,161,705)	$74,889	$1,181,082
After-tax IRR	–12.21%	15.55%	22.10%

The most likely scenario shows loans of $15 million and grants and land write-downs that total $6 million. We are prepared to put in slightly over $2 million in equity. Once again, major technology companies wrote down the total develop-ment costs to $23 million from $28 million. The $5 million was considered a gift to the community for their support of start-up and more established technology companies.

This scenario is based on preliminary commitments from foundations so that the high construction costs will be minimized. The equity is considered a combina-tion of contributed capital from the Alliance plus equity from foundations. Although the returns are low compared to those from unsubsidized projects, they are very reasonable and acceptable for a catalyst project.

Costs and Benefits

Community Courtyard has the potential to be a vital catalyst project for all of Mid-dle Town. However, there are some inherent and external risks associated with the development:

- Economic downturn—high vacancy rates, lower rents
- Unexpected construction costs
- Inability to provide equity
- Failure to qualify for grants or low-interest loans

- Negative perception of the neighborhood
- Stakeholder opposition

Although the potential risks are many, the direct and indirect benefits to the community outweigh any of the risks. These benefits include

- Community reinvestment
- Increased retail services
- Low- and moderate-income housing
- Community services
- 130 new jobs
- Social integration of different ethnic and religious groups
- Creation of an architectural landmark

Project Benefits/Risks

	Community Alliance	City of Middle Town	Equity Provider	Lender	Grant Provider
Risks	Failure to provide cash flow	Bond debt default	Loss of equity	Default on loan	Failure of project
Benefits	Quality of life	Increased tax revenue	High return on investment	Community reinvestment and loan interest	Community reinvestment

Conclusions and Recommendations

Our group is specifically seeking to raise additional sources of funding from foundations, the city, and the state for outright grants, tax increment financing, land write-downs, and other low-interest loans. We will need subsidies that total close to $29 million.

We believe in this project because it will improve the quality of the lives of the community, provide increased jobs, and bring in increased revenues for the citizens and Middle Town.

The large amount of grant fund-raising needed for the projects represents a difficult endeavor. The nature of the project will reduce the effort required to raise these funds. This is an important project for the community and neighboring towns. For that reason, we request that you give our business plan serious consideration.

Conclusion

Local economic development finance is more than the deal. It is concerned with the development of real wealth within large and small communities. It is concerned about low- and moderate-income residents who need sustaining jobs and economic prosperity. In this book, we have focused on the combination of tools and processes to ensure this outcome.

Though we are very aware that no one strategy guarantees the achievement of social or economic equality or improvement in the quality of lives within a community, we have demonstrated the range of possibilities. Many finance tools can be used for the development of new businesses or projects. Human services, new real estate development, and employment in advanced technology businesses are real opportunities for communities. Understanding the tools and knowing where the decision makers are allows all communities to compete for improved economic standards.

We believe that the strategic business plan enables leaders to compete in the marketplace for funds to implement local economic development. In preparing the strategic business plan, the local economic development professionals, the community development corporation leaders, the public policy executives, and the entrepreneurs have the ultimate responsibility to communicate their goals to the financial community. The strength of the business plan is that it "speaks the language" of the financial executives, foundation boards, and lenders who are evaluating a development project, a technology venture, or a child care facility. The critical task, then, is to ensure that an adequate financial plan is well documented. Most importantly, the organizational structure must be sound because people do business with people. Lenders must be confident that they are lending to those business leaders or entrepreneurs who have the expertise to execute the business plan and repay the loans.

We cannot overemphasize the importance of sound local economic development planning. This planning begins with the philosophy for the development and not necessarily the opportunity to make a profit. Therefore, the real social and economic goals must be determined and well articulated before the process commences. The type of organizational form is important as well. Some development forms are better suited to certain businesses or organizational forms than others.

The business plan approach provides the correct template for most local economic development projects. The use of a business plan approach should be interpreted, not as a "business-first and people-second" approach, but as a continuation of a rigorous planning paradigm. The way you apply the tools provided in this book will vary with your goals and your resources. Creativity and innovation are essential. Push the boundaries to explore what is really possible. Use all your available resources and you will surely achieve your goals.

Appendix A shows a complete business plan for a local economic development project. This plan can serve as a guide for developing local economic development strategic business plans.

Sample Business Plan:
Embassy Townhomes

Contents

III. Marketing Plan and Overview

    A. Marketing Plan
    B. Residential Market Overview
        Park Mile
        Windsor Village
        Hancock Park and Fremont Place
    C. Commercial Office Market Overview
    D. Demographic Overview

IV. Organizational Structure and Company

    A. Structure
    B. Company Profile
    C. PMG Principals
    D. PMG's Profit and Loss Statement

V. Benefits and Risks

I. Executive Summary

The 2.05-acre project site is located at 4375 Wilshire Boulevard at the center of the exclusive Park Mile District of the Wilshire Corridor. The Park Mile area is located west of the Transit Oriented District that begins at Wilton Place and extends to Normandie, and east of the Corridor's Museum District, which begins at Highland and extends to Fairfax. Neighborhoods directly adjacent to the site include the Hancock Park residential area, which features large, estate-type homes; Fremont Place, an exclusive gated community located directly southwest; and Windsor Village, a community that offers an array of luxury condominiums, townhomes, and apartments.

The site currently houses the Scottish Rite Cathedral, which was closed in 1994 because of high operation costs and limited reuse potential stipulated by the Park Mile Specific Plan. Once a stately building, the vacant, fenced-off cathedral has lost its glory and has become an eyesore to surrounding communities. An initial viability study conducted by the Park Mile Group (PMG) revealed that the site would require immense rehabilitation to comply with current codes. Consequentially, exorbitant rehabilitation costs ruled out the possibility for the cathedral's reuse and conversion.

The best use for the site in the context of Park Mile Specific Plan, existing land use, and long-term market demand has been determined by PMG to be multi-family residential or commercial office use. Market analysis conducted by PMG demonstrates a relatively high vacancy rate for commercial offices (27%) and a low vacancy rate for multifamily residences (5%). Responding to market trends, PMG developed a twofold program that first establishes marketing strategies to promote the Park Mile area as the premier location for the concentration of consulates offices. Complementary international business activities would be supported by funding through a privately sponsored business improvement district (BID) composed of existing Park Mile business and property owners. The second facet of the program proposes a PMG-financed multifamily residential development to be built on the project site to combat increasing housing shortages. This twofold program is translated into the "Luxury Living and World-Class Business" concept, a comprehensive approach to ensure both the short-term and long-term goals of existing Park Mile business and property owners.

PMG's market research further supports the concept. Thirty-five of the 56 consulates located in the city of Los Angeles are currently located along Wilshire Boulevard. A recent survey conducted by PMG found that a primary factor for this type of organization is east-west accessibility. As the main thoroughfare linking the downtown area with West Los Angeles, Wilshire Boulevard has been a prime location for consulates. PMG found that consulates favor PMG's concept of creating a consulate district. They believe that geographic proximity between consulates would help promote common interests and cultural understanding, and they are in favor of a central district that lends itself to both business and comfortable living.

The proposed multifamily residential development consists of 66 luxury condominium townhome units, totaling 107,000 square feet of newly built residential space and occupying two buildings with shared open landscaped grounds and amenities. The two buildings would be developed and sold in two phases to minimize PMG's risk. Four floor plans were designed: twenty-two 1,400-square-foot units, twenty-two 1,600-square-foot units, twenty 1,800-square-foot units, and two 2,500-square-foot units. The price range of these units starts at $300,000 and goes up to $500,000, reflecting existing pricing trends in the Park Mile district. The development would generate a total of $22,880,000 in sales revenue.

The development cost of the project is estimated at $18,786,080 (calculated based on $30 per square foot for land cost, $3 per square foot for demolition, $150 per square foot for construction, $2,500 per stall for above-ground parking, and a flat rate of $60,000 for amenities). Twenty-five percent of the total project cost, or $4.7 million, is developer equity being invested by PMG. Remaining development costs in the amount of $14,089,560 will be financed through a conventional lender at an interest rate of 10% over a period of 2 years. Years 1 and 2 net operating income totals $22.6 million, and cash flow after taxes is equal to $10.3 million in Year 1 and $10.7 million in Year 2, with cash-on-cash ratios of 59% and 62%, respectively. According to PMG's estimates, the proposed development will be able to cover its debt service with a debt coverage ratio of 13.21% at the end of Year 1 and 13.72% at the end of Year 2 and will generate a 38% profit based on the initial equity investment.

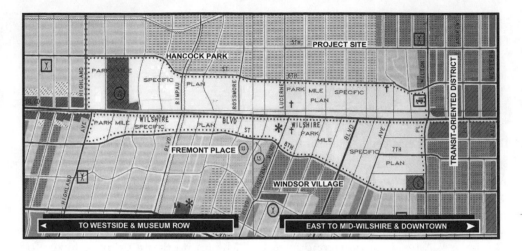

II. Description of Proposed Development

▬ A. Project Concept

Embassy Townhomes is a 66-unit residential housing development that represents one strategic component in a larger, long-term "Luxury Living and World-Class Business" concept envisioned for the Park Mile area. Specifically, PMG seeks to promote the Wilshire Corridor area located in the Park Mile as a premier location for the concentration of embassy and consulate offices and complementary international business activities and further seeks the continued development of residential townhomes and condominiums.

Primarily single-family residential with limited-use commercial office building space spanning its east-west center along Wilshire Boulevard, the Park Mile area is one of a handful of well-preserved and -maintained residential concentrations located in what is otherwise an extremely dense and highly urbanized central area of Los Angeles. Its proximity to downtown Los Angeles, its link to the Miracle Mile area and other westside centers, and its central location on the Wilshire Corridor make the Park Mile Area an important and strategic component in the continued revitalization of the Wilshire Corridor and downtown Los Angeles. The Park Mile Specific Plan is part of the Larger Wilshire Plan and represents one of the most restrictive and clearly defined specific plans in the city of Los Angeles.

Commercial office space within the Park Mile is best characterized as 50% owner occupied and 50% renter occupied. Vacancy rates within the office rental segment remain high at 20% or more (27% according to one source). Limited reuse and retail potential require a long-term vision that will (a) secure long-term economic health for Park Mile property owners and existing businesses, (b) maintain property values, (c) decrease office vacancy rates, and (d) promote business and development within the narrow limits of the Park Mile Specific Plan. In keeping with the Park Mile Specific Plan, the concept seeks to address these issues while maintaining the goals established by the Park Mile Specific Plan.

Specific Plan goals include the promotion and preservation of the area's character and "parklike setting" and of a "restricted intensity, open and richly landscaped environment, consistent with the residential environment which surrounds it." The purpose of the Park Mile Specific Plan is to protect the low-density, single-family residential nature of the area and to promote only those developments that are compatible with adjoining residential neighborhoods—developments that reinforce the parklike setting and characteristic pattern that provides the area with an image, sense of community and orientation, and visual differentiation from adjoining Wilshire and Miracle Mile centers.

Market research conducted by PMG has identified embassy, consulate, and supporting international business entities as natural tenants for existing office space. The "Luxury Living and World-Class Business" concept envisioned for the Park Mile area is a natural approach supported by market research, which identifies and addresses the needs of existing commercial property owners and supports the long-term objectives of residents and the Specific Plan. Additionally, the concept represents a major catalyst for reinvestment along the eastern Mid-Wilshire, Park Mile, and Miracle Mile portions of the Wilshire Corridor and adjacent residential communities to the east and south of the Park Mile area.

▬ B. Project Site Overview

Located at 4357 Wilshire Boulevard, this 2.05-acre site is on the north side of Wilshire Boulevard between Lucerne and Plymouth Boulevards, directly across the street from the Ebell Theater and the recently rebuilt Korean Christian Church. Together, these three parcels form a distinct triangular gateway. Lucerne Boulevard, which runs north-south through adjacent residential areas, cuts across Wilshire Boulevard at the center of this triangular gateway, offering a unique physical land use pattern that provides a natural link across the traffic-heavy Wilshire Corridor between the Hancock Park and Windsor Square residential communities.

The residential area running north along Lucerne from Wilshire to Beverly Boulevard is characterized by well-maintained mansions and estate single-family properties valued at between $750,000 and $2.5 million. The residential area extending south along Lucerne is characterized by a three-square-block cluster of condominiums, townhomes, and luxury apartments; large, mansion-style homes; and a residential area of smaller, one-story, single-family homes that extends south to Pico Boulevard. Just two blocks south of Wilshire, at the heart of where

these three residential clusters come together, is a well-maintained neighborhood park that boasts a playground and picnic area.

The project site currently houses the Scottish Rite Cathedral. Built in 1960, the cathedral is an elaborate special-use structure designed for club/lodge use. Measuring four stories and approximately 160,000 gross square feet of building space, the structure and site are privately held by a nonprofit social organization. The cathedral was boarded up in 1994 as a result of high operating and maintenance costs. Restrictive zoning and land use regulations stipulated by the Park Mile Plan severely limit the reuse potential for the structure.

A viability study was conducted by PMG to determine the feasibility for rehabilitation, reuse, and conversion of the existing structure. Though market interest for reuse of the site was determined to be high, the exorbitant cost of rehabilitation, continued maintenance, and restrictive zoning proved to be major constraints, making reuse and conversion unrealistic. Since its closure, the building interior and exterior have both fallen into ill repair, and currently the site is an eyesore in what otherwise is a well-preserved and exclusive section of the Wilshire Corridor. The cathedral was put on the market in 1994 at an asking price of $11 million. Subsequently, the asking price has been reduced to $6,000,000. The highest and best use for the site in the context of Park Mile Plan specifications, existing land use, and long-term market demand has been determined by the PMG group to be both multifamily residential and/or commercial office use. Market studies conducted by PMG have further determined residential multifamily development to be the best and highest use for the site over the short and long term, given existing and projected market trends for the area.

The Embassy Townhomes development presented here is based on PMG's feasibility and market analysis. Land prices/costs for the property have been determined by market analysis and project feasibility and not by the present owners' asking price. An offer for the site of $20 per square foot ($1,783,580) will be made based on PMG's findings. Because of its unique triangular location and proximity to adjacent features and neighborhoods previously discussed, this site has been determined to be ideal for the townhome development. However, should the purchase price for this site become an obstacle, four sites comparable in size and located within the Park Mile that have never been developed have been identified and will be considered as secondary site locations.

▬ C. Development Specification

The proposed residential development consists of 66 luxury condominium townhome units, totaling 107,000 square feet of newly built residential space and occupying two buildings with shared open landscaped grounds and amenities. Amenities for the development include a pool, a spa, a clubhouse, and shared lushly landscaped garden areas. The existing 2.05-acre property will be modified to accommodate these luxury units for sales and operation within a 2-year time period. This section is a discussion of the project specifics and an analysis of the fiscal viability of the project.

Unit Floor Plans

The project is currently designed to accommodate 66 luxury units, supporting four different floor plans. The first model is a 1,400-square-foot two-bedroom/bath unit with a small workspace/den. The second unit type is a 1,600-square-foot two-bedroom/bath unit that has a larger den area as well as a minioffice bay with multiple phone and computer connections. The third unit type is an 1,800-square-foot three-bedroom/bath unit with a den. Together, these three floor plans make up 64 units of the planned 66-unit development. Additionally, two "penthouse-type" units with a luxurious size of 2,500 square feet of space have been designed, completing the remaining two units of the 66-unit development. These units offer three bedrooms, three baths, a large den, and a private rooftop patio. Association fees have been calculated at $250 per month for the smaller 44 units and $350 per month for the larger 22 units.

Unit Pricing

Market research conducted by PMG supports the following pricing structure. A detailed discussion of the housing market in the Park Mile is discussed further in this document. All units are priced at the high end and reflect pricing trends in the Park Mile.

Table A.1

Unit Floor Plans	Sale Price ($)	Unit Sq. Ft.	Total Units	Unit Assoc. Fee ($)	Total Sales Units ($)
2-bdrm, 2-bath, den	300,000	1,400	22	250	6,600,000
2-bdrm, 2 bath, den	340,000	1,600	22	250	7,480,000
3-bdrm, 3-bath, den/office	390,000	1,800	20	350	7,800,000
3-bdrm, 3-bath, den, office, rooftop garden	500,000	2,500	2	350	1,000,000
Project Totals	**N/A**	**107,000**	**66**	**18,700/mo.**	**22,888,000**

Development Costs

Development for the project will be conducted in two phases, with one building (33 units) being completed in Phase and Year 1 and a second 33-unit building being completed in Phase and Year 2. Total project costs for Years 0 to 2 are $18,786,080. Preconstruction costs include the demolition of an existing 160,000-square-foot structure on the project site, estimated at $480,000, and land costs, calculated at $20 per square foot. Construction costs, including both hard and soft costs, are $150 per square foot, with a total buildable space of 107,000 square feet. Parking will be provided at 2.5 spaces per unit at a cost of $2,500 per space, which also provides for guest parking. Amenities including a pool, a spa, a clubhouse, and shared gardens and landscaped space are estimated at $60,000.

Table A.2

Development Costs	Unit Cost	Total Units
Land cost	$20/sq. ft.	$1,783,580
Demolition	$3/sq. ft.	$480,000
Construction	$150/sq. ft.	$16,050,000
Parking (165 spaces)	$2,500 ea	$412,500
Amenities	—	$60,000
Project Total		**$18,786,080**

Financing Assumptions

Twenty-five percent ($4.7 million) of the total project cost is developers' equity being invested by the Park Mile Group. Remaining development costs in the amount of $14,089,560 will be financed through a conventional lender at an interest rate of 10% over a period of 2 years. Development costs and profits have been calculated on the basis of a single development loan and on a 50% sales of units during Year 1 and a balance of sales in Year 2. However, a variety of other financing options remain to be explored, such as phase financing, whereby Year 1 units are financed and sold with revenues used to finance Year 2 units. Another option discussed in greater detail in the marketing plan is an implementation strategy for unit presales.

Year 1 and 2 operating income totals $22.6 million, and cash flow after taxes is equal to $10.3 million in Year 1 and $10.7 million in Year 2 (totaling $21 million). Proceeds from sales of the units in Year 1 will equal $3.0 million, and proceeds from sales in Year 2 will equal $2.7 million. Cash flow after taxes for Year 1 will be close to $7.2 million and will be $7.5 million for Year 2. An indicator of return that makes a determination on the viability of this project is the cash-on-cash ratio. For Year 1, the return is 59%, and for Year 2, the return increases slightly to 62%.

Table A.3

Financing Assumption		Year 1	Year 2	Total/Avg.
Total project cost	$18,786,080			
Equity (25%)	$4,696,520			
Loan	$14,089,560			
NOI		$11,151,800	$11,578,807	$22,730,607
Debt coverage ratio		13.21%	13.72%	13.47%
Cash-on-cash		59%	62%	61%
Cash flow after taxes		$7,275,208	$7,555,335	$14,830,543
Sales proceeds		$3,014,936	$2,794,079	$5,809,015
IRR (before sales)		0.35	1.00	0.68
IRR (after sales)		1.19	1.20	1.20

According to PMG's estimates, the proposed development will be able to cover its debt service obligations, with a debt coverage ratio of 13.21 during Year 1 and 13.72 during Year 2. Typical debt coverage ratios range from 1.25 to 1.50 for development projects. Overall, according to PMG estimates, this project will generate close to 38% profit, based on its initial equity investment of $4.7 million. The likely profits will total $1.17 million for the 2 years of this residential project.

III. Marketing Plan and Overview

A. Marketing Plan

The Embassy Townhomes marketing strategy is an important element to the overall success of the Park Mile development project. The "Luxury Living and World-Class Business" concept was developed as a comprehensive approach to ensure both the short-term goals of the Embassy Townhome Development and the long-term goals of existing Park Mile business and property owners. The concept and marketing strategy is designed to (a) secure long-term economic health for Park Mile property owners and existing businesses, (b) maintain and increase property values, (c) decrease office vacancy rates, and (d) promote business and development within the narrow limits of the Park Mile Plan.

The Embassy Townhomes' multifaceted concept focuses on (a) retaining the unique characteristics of this prime, upscale, and centrally located urban community of primarily single-family and condominium homes and (b) promoting the strategic Wilshire Corridor portion of the Park Mile as a premier location for the concentration of embassy and consultant offices and complementary international trade activities. The corridor's strategic central location and proximity to both downtown and westside centers make this a natural location for these organizations. Additionally, the lush, parklike atmosphere, low traffic/congestion, and proximity to exclusive residential areas, including Hancock Park, Fremont Place, and Windsor Village, make this an ideal location in which to live.

PMG's market research further supports the concept. Thirty-five of the 56 consulates located in the city of Los Angeles are currently located along Wilshire Boulevard, clustered in three areas: Westwood (to the west), Miracle Mile and Park Mile (central), and the Wilshire District (to the east). Often, these are several unrelated consulates located in a single office building. A recent survey conducted by PMG found that a primary factor for these types of organizations is east-west accessibility. Wilshire has long been a main thoroughfare linking the downtown area with West Los Angeles and has long been a prime locational choice for consulates. PMG found that consulates favor PMG's concept of creating a consulate district. They believe that geographic proximity between consulates would help promote common interests and cultural understanding, and they are in favor of a central district that lends itself to both business and comfortable living.

Some of the primary problems these types of organizations are experiencing in their present locations are increasing rental rates, inadequate and expensive parking, and—for those presently located along the eastern-most portion of Wilshire—heavy traffic congestion, inconsistent quality of property and street-area maintenance, and perceptions of crime, which make the area less visitor-friendly. Existing quality vacant office space, a centralized location, and the parklike amenities in the Park Mile offer such organizations an optimal choice for relocation.

Another important marketing component is the planned formation of a Park Mile Association (PMA), which would be established in Year 1 of project development. Operating much like a BID, this association, whose members would be Park Mile business and property owners, would self-assess in the amount of $25 per linear square foot of member-owned property fronting Wilshire Boulevard per year. Potential revenues generated for the marketing and promotion of the Park Mile concept are $198,000 (75% member participation rate) in the first year. Funds generated will be specifically applied to the following activities:

1. Developing marketing tools to target potential consulate, embassy, and complementary international business tenants for existing surplus office space, which is currently at between 20% and 27%

2. Developing marketing tools to target potential individual, corporate, and organizational occupants for existing planned and potential residential units, including condominium/townhome and rental

3. Providing administrative, legal, and other such support (on a project-by-project basis as approved by members) for PMA-sponsored projects and developments where community consensus building, permitting, and/or other such approvals are required

4. Initiating investment interest and activities in member-approved PMA-sponsored residential and commercial development projects

Additionally, surplus PMA funds collected would be placed in a venture capital investment pool and used to seed sponsored residential and commercial development and redevelopment within the Park Mile. Funds would be made available to members and nonmembers at market or below-market interest rates on a project-by-project basis as approved by member-voted project committee members. Though PMG has worked to develop interest and to organize the PMA for the Park Mile, PMG's long-term role will be as a participating member.

▬ B. Residential Market Overview

Residential properties in and adjacent to the Park Mile vary in size and price. There are four adjacent and distinct residential nodes: Park Mile, Hancock Park, Fremont Place, and Windsor Village.

Park Mile

Five luxury condominium complexes, one Tudor-style townhome complex, and one apartment building can be found on Wilshire Boulevard in the Park Mile adjacent to low-rise (three- to four-story) office buildings. Condo and townhome complexes range from 22 to 70 units and from two to four stories in height. Some

have subterranean parking and others have grade-level parking with units above. Most are sized between 1,500 and 1,900 square feet and feature two bedrooms, two bathrooms, a den, 24-hour security, and amenities such as a spa and pool. Prices for these condominiums and townhomes range from between $265,000 to $389,000. Lease rates average $2,500 per month. The single apartment building located on Wilshire is a historic landmark currently undergoing restoration. This eight-story building has a waiting list of anxious renters, and rents are expected to be in the $1,100 to $1,500 range.

Windsor Village

Located just south of Wilshire Boulevard, Windsor Village features three concentrations of housing types. The area beginning at Lucerne and running east to Wilton Place offers a mix of condominiums, townhomes, and apartments. Prices here range from $169,000 to $220,000, and rents run between $1,100 and $1,600 per month, or between $1.00 and $1.20 per square foot depending on the age of the structure. A concentration of stately homes can also be found in this area, with prices ranging from $550,000 to $1,500,000. Also beginning in the Windsor Village area and running south to Pico is a large concentration of well-kept, one-story,

single-family homes averaging 1,200 to 1,500 square feet and having prices between $250,000 and $350,000.

Hancock Park and Fremont Place

Hancock Park is a fairly large geographic area that begins at Wilshire and Wilton to the southeast and extends to Beverly and LaBrea to the northwest. The portion adjacent to the Park Mile area is characterized by stately mansions and estates. Fremont Place is a highly protected gated community. Secluded neatly behind condominiums and office buildings on the south side of Wilshire Boulevard at around Fremont and Rossmore, this area extends to Olympic Boulevard to the south and features an estimated 114 homes. Property values in both of these exclusive areas range from between $850,000 to $2,800,000.

— C. Commercial Office Market Overview

Fifty percent of the commercial office space within the Park Mile is owner occupied, and 50% is renter occupied. Vacancy rates are currently at 27%, though area brokers expect rising rates in popular westside locations to have an impact during the next year as rent-sensitive tenants begin seeking more affordable locations, including the Park Mile. Much of the vacant office space can be found in new (3- to 10-year-old) buildings, and parking in most commercial buildings is free. Rental rates in the area vary from $1.10 to $2.10 per square foot and average at about $1.50. Some owner-occupied buildings are considerably higher, at more than $3.00 per square foot. Land prices in the Park Mile vary from between $20.00 and $35.00 per square foot. Unlike immediately adjacent areas—Wilshire Corridor east of Wilton and the Miracle Mile area immediately to the west—the large number of owner-occupied properties has had a stabilizing effect on the Park Mile area, which has resulted in stable rental rates and consistent property maintenance throughout the area. Locational benefits cited by tenants include the absence of traffic and congestion and surplus parking. The price tenants pay for these benefits includes the lack of supporting amenities such as restaurants and personal services (disallowed by the Specific Plan), which employees can access only by driving out of the immediate area.

Existing owner-occupier tenants and commercial renters include a host of insurance companies like Farmers (which occupies three large buildings), Continental, Massachusetts Mutual, Mercury, and Japan Life. Other tenants include the Virgin Entertainment Corporate Offices, the Continental Graphics Corporate Office, the Arthritis Foundation, Pitney-Bowes, the Korea Times, FM Seoul, the

Korea Trade Center, the Seoul Information Center L.A., Japan Life Magazine, the Broadway Federal Bank, and the Ebell Theater. Institutional tenants include St. James Church, Japanese Christian Temple, John Boroughs Jr. High School, the Theater Arts Center, and a private day care and grade school.

▬ D. Demographic Overview

Finding: Park Mile households are significantly wealthier than other area and city residents.

The Park Mile area is composed of three census tracts: 2110, 2127, and 2117.01. Though they share many demographic similarities with the larger Wilshire area and with the city as a whole, the Park Mile is surprisingly unique in many regards. Census data from 1989 demonstrate that average household income in the Park Mile is significantly higher, at $77,553, than income in the larger Wilshire area and citywide, where averages are $38,244 and $45,701, respectively. Median family incomes average $134,646 in some parts of the area, indicating a fairly affluent population.

Finding: Whites and a growing Asian population account for more than two thirds of all Park Mile residents.

Historically, Park Mile has been a white community, and whites account for 45% of the population, but demographic shifts indicate a growing Asian

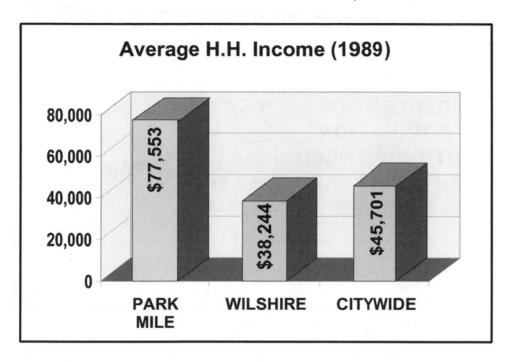

population—26%—which is actually greater than in the Wilshire area, where Asians constitute 21% of the population. Hispanics are also well represented at 15%, compared to 39% in the larger Wilshire area. African Americans account for 6.4% of the population in the area, and other, nonspecified races account for 7.6%.

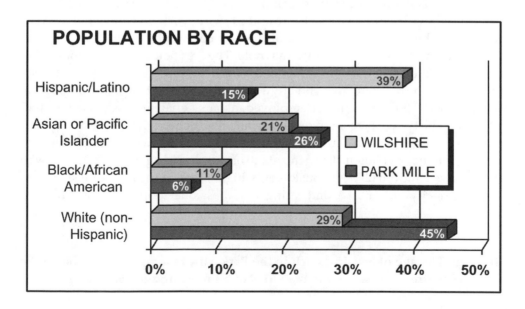

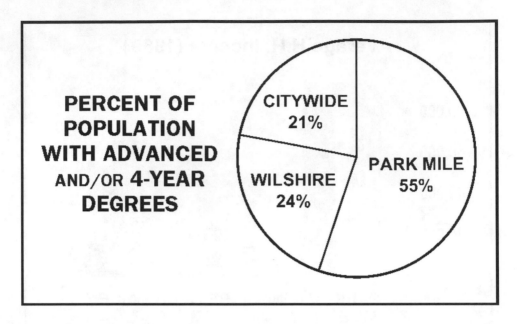

Finding: Park Mile residents are significantly more educated than other Wilshire area and city residents.

A strikingly large number of college-educated persons reside here. Sixty-two percent of the Park Mile population have completed an advanced college degree and/or a 4-year college education, compared to 26.5% for the larger Wilshire area and 23% citywide. Much as in Wilshire and citywide, Park Mile shows 28% of residents to be between the ages of 22 and 39, as compared to 39% for the larger Wilshire area and 35% for the city as a whole. The area houses a larger population of residents aged 40 to 50, with 26% of residents in the age group compared to 19% and 20% for the Wilshire area and citywide, indicating a stable resident population. Almost 75% of residents are employed in managerial, professional, or technical fields. Thirty-two percent of residents here live alone, indicating a strong market for condominium and townhome residential space. Twenty-six percent are married and have children. The percentage of couples living in the Park Mile who do not have children (27%) is considerably higher than in the larger Wilshire area (19%) and citywide (22%), indicating a professional population of working couples.

Finding: The age of housing stock in the Park Mile is comparable to that of the larger Wilshire area, with more than 51% of all housing aged 50 years or more.

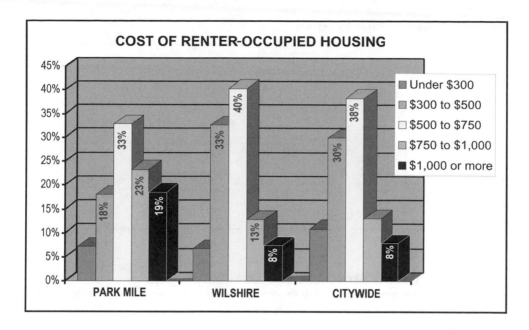

Newer housing stock is growing faster in the area, however, with more than 10% of all housing less than 10 years old and 10% between 10 and 19 years old, as compared to only 4% and 6% in the Wilshire area, indicating a steady market demand for housing.

Finding: New housing construction indicates a steady market for housing.

Fifty-six percent of all owner-occupied housing in the area is valued at $500,000 or more, compared to 29% in the Wilshire area and 15% citywide. Half of the population here spend less than 20% of their income on housing, and slightly more than half reside in owner-occupied, multiple-family dwellings. Renter-occupied dwellings account for 47% of all residential units, with more than 41% of all renters spending 30% or more of their income on rent, compared to 52% in the larger Wilshire area. Almost 20% of Park Mile renters pay more than $1,000 (based on 1989 census figures, actual rental rates are presently much higher) rent, compared to only 7.5% in the Wilshire area and 8% citywide. Of the total Park Mile population, 60% spend between 5 and 34 minutes commuting to work, and 20% spend between 5 and 14 minutes, indicating that the area's central location and proximity to both east and west urban center make it a definite draw for current and potential residents.

Finding: The Park Mile area can support luxury multifamily housing.

Park Mile Organizational Structure

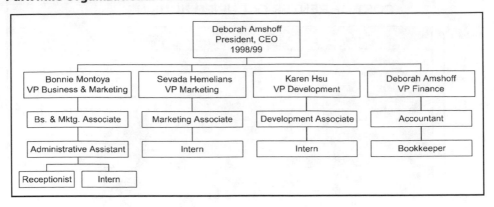

IV. Organizational Structure and Company

▬ A. Structure

The Park Mile Group is a privately owned and operated firm managed wholly by the firm's four founding principals. Each year, one principal is elected to serve as PMG's president and chief executive officer for a 12-month period. Though most decisions are made by all four of the firm's founding principals, situations occasionally arise where all principals are not available. In such cases, during his or her term, the elected principal is responsible for making judgment calls and overseeing the best interests of the firm and its remaining founding principals, associates, support staff, and PMG clients. During his or her term, the acting CEO is also responsible for scrutinizing, on a monthly basis, all financial expenses to ensure the firm's financial stability and long-term health. Additionally, he or she is responsible for representing the firm at public and private functions.

All revenue-generating activities conducted by PMG and its founding partners are reinvested into the firm's holdings. PMG development projects are funded through the firm's holdings. All reinvestment and profit distribution are determined by consensus and distributed equally among PMG principals. Where development projects are 100% PMG, project financing, administration, and management requirements are supported by existing PMG staff and resources. In the case of joint venture development activities with other public and private organizations, PMG becomes partner, and project-specific contracts outlining each partner's responsibilities are drawn up. In partnerships where additional administrative and management resources are required, overhead expenses are calculated and distributed by the joint partnership. The Embassy Townhomes development is a 100% PMG project. The Park Mile Association (PMA) established by PMG will be administered by an independent, elected board of directors, and PMG will act only as a PMA member.

━ B. Company Profile

Founded in 1992, the Park Mile Group (PMG) is a consulting, marketing, and real estate development firm. It provides a variety of specialized consulting services to public and private sector economic development organizations, developers, and business and community associations. PMG's specialized services include

- Market research
- Strategic planning
- Grant preparation
- Project feasibility analysis
- Business attraction and retention
- Real estate analysis
- Real estate development
- Project implementation

PMG's development activities have expanded in recent years, and past PMG development undertakings have included multifamily residential, commercial, retail, and mixed-use rehabilitation projects. All PMG development projects are selected and managed by PMG's founding principals, who also serve as the primary equity investors for most PMG development undertakings. However, recent PMG development successes have captured the interest of outside investors interested in project-specific partnerships with PMG.

Each development project considered is sponsored by one or more of PMG's principals who have a special interest in seeing a project completed. All projects are guided by the principals' shared philosophy, which calls for projects to have a positive, long-term social and/or economic impact on the community in which they are built. Projects must be self-sustaining and developed within the framework of the needs and values of the community. PMG especially seeks to participate in development projects that promote urban renewal and improved quality of life. Many of PMG's developments serve as catalysts to increase economic development within a given community. PMG is known for its attention to detail, quality site planning, and design, as well as for its ability to inspire community vision and capture community commitment.

PMG's professional staff possess a broad range of experience in all aspects of economic development, real estate planning, and marketing. Combined, the firm's principals have more than 40 years of professional experience in providing planning services to public and private sector clients. The diverse, creative, and practical combination of skills drawn from each of the firm's principals and professional staff enables PMG to approach a broad range of complex issues and to develop comprehensive strategies and solutions that ensure a strong foundation and long-term success. The firm is managed by its four founding principals and employs a staff of 10 planning, real estate, business, and administrative professionals.

■ C. PMG's Profit and Loss Statement

The following profit and loss statement is designed to provide an overview of PMG's financial condition during the most recent fiscal year.

Table A.4 The Park Mile Group For-Profit
Profit/Loss Statement, as of December 31, 2000 (in $)

Sales and cash reserves	17,500,000
Cost of operations	1,800,000
Gross profit	**15,700,000**
OPERATING EXPENSES	
Wages	500,000
Executive compensation	300,000
Payroll taxes	210,000
Miscellaneous	190,000
Material and supplies	55,000
Rent	64,000
Advertising	61,000
Interest	156,000
Telephone and utilities	80,000
Insurance	63,000
Accounting and legal	85,000
Total	**1,764,000**
PROFIT/LOSS	
Operating income	17,500,000
Operating expenses	1,800,000
Net operating income	**15,700,000**

V. Benefits and Risks

In development of the proposed Embassy Townhomes project, two primary stake-holders are involved with project finances. The developers, PMG, and the project's conventional lender would assume the greatest financial risks and anticipated benefits. To a much lesser degree, commercial property owners located along Wilshire Corridor in the Park Mile District would also assume financial risk through their involvement in the planned formation of a Park Mile Association. Operating much like a government-sponsored BID, PMA members would self-assess funding for collective marketing and promotion of the Park Mile commercial corridor. Finally, four additional stakeholder groups would be affected by the proposed multifamily development, each carrying distinct levels of benefit and risk but no financial investment.

As summarized in the following matrix, PMG would assume the greatest financial risk. PMG also has the greatest opportunity to benefit, both financially and personally, from success of the Embassy Townhomes project and development of the overall "Luxury Living and World-Class Business" concept. During development, PMG principals would maintain decision-making control concerning their company's assets, annual budget, site development, marketing, profit distribution, and benefits such as tax write-offs. However, PMG would expect to pay the project conventional lender first, even in the event of cash flow problems, and would adjust its annual budget to ensure that its debts were covered before calculating its return on investment.

The conventional lender, most likely a commercial bank that had provided development loans to PMG in the past, would maintain control of the loan money through a negotiated loan contract. In return for lending money, they would achieve 10% compounded annual interest on the loan amount, whether the project is barely profitable or wildly successful.

Commercial property owners in the Park Mile District, as prospective PMA members, are likely to gain the most while realizing relatively low risk. At $25 per linear foot of Wilshire Boulevard frontage, an office building owner with 200 feet of frontage would provide $5,000 annually toward joint marketing and promotion of Park Mile commercial rentals. This could be money diverted from individual marketing budgets. Potential benefits include lower vacancy rates and, eventually, the opportunity for higher rents.

Embassies, consulates, and international trade associations have the potential to achieve high levels of positive synergy by locating in the Park Mile District. They might also realize lower rents than those to be paid in their current locations because office rents throughout the region are rebounding from recession lows. Many office renters are moving rather than signing contracts for rents that are 25% or higher than just 2 years ago. These international organizations would also gain the opportunity to live and work in the centralized and desirable Park Mile District.

Park Mile District residents may finally realize implementation of the Park Mile Specific Plan, with a concept that goes beyond a single-site development. Even if Specific Plan restrictions are met, however, nearby homeowners may experience temporary construction impacts as well as longer term impacts related to traffic, noise, and parking.

City of Los Angeles decision makers, such as Councilman Ferraro, will need to make a political decision to support (or not) the proposed development and its concept of a consulate center. Assuming the proposed development is accepted, city decision makers have the risk of looking bad if implementation does not go through as anticipated.

Overall, PMG has minimized financial and political risks by preparing conservative pro formas and gaining local neighborhood support before officially applying for development approvals. Other stakeholders would also be likely to be conservative until the project seemed to be "a sure bet." If this is a successful project, as anticipated by PMG, it will act as a catalyst to spur greater development and associated benefits for all the stakeholders.

Table A.5 Benefit and Risk Matrix

Stakeholders	PMG Developer	Conventional Lender	Park Mile District Commercial Property Owners/ BID Members	Embassies, Consulates, International Trade Associations	Park Mile District Residents	City of L.A. (Councilman Ferraro)
Investment	Equity investment for 25% of total project development costs	Loan for 75% of total project development costs	Annual self-assessment of $25 / linear foot along Wilshire Blvd.			
Return (%)	62% cash-on-cash return; 1.19 IRR (including sales)	10% interest				
Cash flow	$21M after-tax cash flow, with $1.17M profit	$693,131 on 24-month $14,089,560 loan				
Benefits	Profit; long-term economic health of Park Mile District; decreased office vacancy rates; increased business development; creation of an area that lives up to slogan "The Place for Luxury Living and World-Class Business"	Profit	Decreased office vacancy rates; long-term economic health of area; increased potential for development of vacant properties; improved image for district	Synergy from close proximity of many international organizations; centrally located consulate district to provide both business and housing; improved commercial image along Wilshire	Loss of neighborhood eyesore; improved commercial image for Park Mile District; concept fits with Park Mile Specific Plan	Increased tax revenues on newly developed properties; concept fits with city plans for Wilshire Corridor
Risks	Market risks: cash flow, return on investment	Market risks: cash flow, return on investment	Potential loss of assessment funds	Potential for cultural conflicts	Potential for increased local traffic, noise and parking issues	Potential for concept not being realized (so appears to fail)

SOURCE: Park Mile Group, Inc.

B

Glossary of Terms

Absorption schedule: The rate or schedule at which properties are sold or leased over a given period of time in a given market or submarket.

After-tax cash flow: The amount of cash returned to the equity investor(s) after deducting debt service and tax liability (or adding tax savings) from the net operating income.

Amenities: Benefits derived from property ownership or occupancy of a building. These are noncash benefits and typically take the form of tangible or nontangible goods or services.

Amortization: A reduction of debt through periodic payments of principal and interest in a defined payment plan.

Anchor tenant: The most important tenant in a development project, whose lease is usually instrumental in securing financing for a commercial undertaking.

Appraisal: An opinion of value that is quantified by one who is qualified to given such an opinion.

Articles of corporation: A legal document required by and filed with the state government where the corporation is chartered. This document describes the purposes for which the corporation is formed and how it will be organized.

Assemblage: Multiple parcels of land joined under common ownership.

Assets: Tangible items of value owned by a business, such as money, merchandise, property, physical buildings, and equipment.

Balance sheet: A financial statement that shows the assets, liabilities, and owner's equity of a business as of a specific date.

Basic employment: Employment associated with business activities that provide services primarily outside the area via the sale of goods and services but whose revenue is directed to the local area in the form of wages and payments to local suppliers.

Bond rating: An estimate of the credit-worthiness of bonds issued by a governmental unit or corporation.

Bonds: Interest-bearing certificates of debt issued by a municipal governmental body or a private corporation to finance physical improvements.

Break-even ratio: Ratio of debt service and operating income to gross income.

Building code: A system of uniform building regulations within a municipality or state established by ordinance or law.

Building society: A financial institution that specializes in providing long-term housing mortgages rather than short-term business and personal loans.

Business development grants: Funds provided to economic development groups by local or state governments that carry no obligation for repayment.

Business risk: The potential loss that can occur through management's control of operations and profits.

Bylaws: Rules, regulations, and controls set by the board of a corporation for the conduct of its business.

Capital: Money or assets invested for wealth creation.

Capital costs: The costs a business pays for major physical improvements such as buildings, equipment, and machinery.

Capital gains: The gain from the sale of an asset. The difference between the total amount received and the total costs.

Capitalization: The process of estimating value through the discount of stabilized net operating income.

Capitalization rate: The rate used to measure an estimate of value through the discount of future stabilized income. It is the rate that lenders and investors expect, given inflation, risk, and the inability to use the investment amount in alternative investments.

Capitalize: To supply a project or business with funds invested by the owners or developers as distinct from borrowed funds.

Cash flow: Spendable income from an investment.

Central business district (CBD):   The business area of a city or town.

Code enforcement: The power of a municipality or agency of local government to require that all properties meet certain standards of construction, maintenance, health, and safety. If a property falls below the minimum requirements and the owner does not satisfactorily repair the property, it can be declared a public nuisance and condemned.

Commercial bank: A financial institution whose primary function is to finance the production, distribution, and sale of goods or services.

Commercial bank loans: Short- or long-term debt issued to businesses or individuals.

Commercial real estate: Real estate buildings held by ownership for the income derived by leases for offices, retail shops, shopping centers, industrial parks, and other business uses.

Commission: A lawfully authorized group of citizens that performs certain tasks or duties in the public interest. An "economic development commission" is a good example of citizens who have been appointed to the task of developing strategies to improve the local economic base.

Community-based organization: A group representative of a significant part of a community/neighborhood that provides services that focus on community development.

Community development corporation (CDC): Usually a nonprofit organization controlled by residents of low- to moderate-income areas to help stimulate economic and physical improvement of the community.

Comprehensive approach: A means of viewing a complicated project as a total picture and then identifying all of the parts that will be needed to complete the proposed project.

Condemnation: The taking of property from an owner by a public agency. The determination by a public authority or agency that a certain property is unfit for use.

Construction loan: A short-term loan to provide financing to develop improvements to a property. The construction loan typically runs from 6 months to 2 or 3 years.

Construction period interest: Loan interest paid or accrued during the construction period.

Contract: A written or oral agreement made between two or more parties to a transaction. In real estate, a contract is usually in writing.

Contractor: A contractor is a person or organization performing certain tasks according to the specific terms of a written agreement.

Coordinated effort: The process of actually working with various levels of government, business, neighborhood organizations, unions, and residents with the intention of establishing who is responsible for certain parts of a project.

Corporation: A legal entity created to perform a function, business, or service. A corporation can enter into a contract without making its owners/executives personally liable.

Cost: The price paid for a good or service.

Cost-benefit relationship: The relationship between what an item costs and what benefit it produces. In economics, this measures the worth of an item.

Credit rating: A rating or evaluation made by a credit reporting company based on a person's present financial condition and past credit history. Credit ratings are used to finance revenue bonds or general obligation bonds for the development of real estate projects. These credit ratings are typically provided by Standard & Poors or Moody's.

Debt: An amount of money owed.

Debt coverage ratio: The net operating income divided by the debt service. A high debt coverage ratio indicates a lower risk for the lender.

Debt limit: A legal restriction on the amount of funds a city can borrow. This amount is normally a certain percentage of the assessed valuation of taxable property in the city.

Debt service: The amount of money required to make regular payments of principal and interest on a loan.

Deed: A written document by which ownership of property is transferred from the seller (called the grantor) to the buyer (called the grantee).

Deed of trust: A three-party instrument between the borrower, a trustee, and the beneficiary or lender for security for a debt.

Default reserve account: A special account created by public and quasi-public agencies in which a financial institution agrees to make high-risk loans and draw upon this default account to recover possible losses.

Demand analysis: A rigorous process to determine the quantity of a good or service that can be sold at a certain price over a stated period of time.

Demographic information: This refers to information about the community in terms of the number of residents in a certain area and their educational level, employment status, occupation, income, age, racial composition, and tenure.

Department of Housing and Urban Development (HUD): A federal department that carries out national housing policies and programs. HUD provides subsidy programs, mortgage insurance, and funding for certain urban projects. HUD also provides loan programs.

Depreciation: A loss or decrease in the value of a piece of property due to age, wear and tear, or deterioration of surrounding properties.

Depreciation (Financial): A deductible expense for building or equipment that reflects the decrease of the useful life of an asset.

Development Authority: An independent agency of local government that possesses special powers beyond those of city government. A public housing authority is a good example because it has the ability to issue special bonds for public housing.

Discounted cash flow (DCF) model: The process by which one estimates the before- and after-tax cash flow and proceeds from an investment, taking it from future value to present value.

Discount rate: The rate used to discount future cash flows to present value. The higher the discount rate, the lower the perceived value.

Disposition: The sale or transfer of property or goods.

Easement: The right to use land owned by another. An example of an easement is a utility right-of-way to allow power lines to cross another's property.

Economically disadvantaged: Any person who is a member of a family that receives public assistance in the form of welfare payments and has a total income that, in relation to family size, is lower than the poverty level determined by the government.

Effective gross income: Gross revenue minus allowance for vacancies and credit loss.

Eminent domain: The right of the government to acquire private property for public use, almost always with adequate compensation to the owner. Also known as *resumption.*

Entrepreneur: One who organizes, manages, and assumes the risks of a business or enterprise.

Equity: The amount of cash or other assets invested in a business.

Equity capital: Funds that owners have personally invested in an enterprise.

Escalation clause: A term in a lease that allows the landlord to pass on certain expenses to the tenant. These expenses typically include increases in operating expenses and real estate taxes.

Escrow: The holding of money and/or documents by a third party until all the conditions of a contract are met.

Facade: The front or principal face of a building.

Fair market value: The worth or value of a property as estimated by a professional property appraiser. It reflects the price at which the property could be sold in a competitive market.

Federal Deposit Insurance Corporation (FDIC):   A federal agency that insures deposits of commercial banks.

Federal Home Loan Mortgage Corporation (FHLMC):   Known as "Freddie Mac," this organization creates a secondary mortgage market for conventional loans and is chartered to circulate funds to poorly capitalized areas. FHLMC also develops new financing instruments to assist in the expansion of the private secondary mortgage market.

Federal Housing Administration (FHA):   A division of HUD that insures loans for within a certain limit.

Federal National Mortgage Association (FNMA):   Known as "Fannie Mae," this organization is a privately owned, government-sponsored agency in the secondary mortgage market. It operates to buy and sell federally insured and guaranteed mortgages and conventional mortgages.

Federal Reserve System: A federal agency that manages the national money supply.

Fiduciary: A person or entity that has a legal duty of trust to another.

Finance charge: A charge one must pay in order to obtain a loan.

Firm commitment: An agreement provided by a lender to make a loan to purchase a particular property. This commitment usually expires after a certain period of time.

Forbearance: The act of delaying legal action to foreclose on a mortgage that is in default.

Foreclosure: The legal process by which a lender forces payment of a loan by legally taking the property from the owner and selling the property to pay off the debt.

Fragmented effort: A noncomprehensive and generally uncoordinated effort to engage in economic development activities.

Franchise: A legal agreement by which a manufacturer or chain store grants the exclusive right to sell merchandise it produces or use the firm's name to sell merchandise.

Front-end costs: Capital required at the early stages of a development project, such as the cost of land and architects' fees.

Gap financing: Financing provided by a second lender, a foundation, or a public agency to bridge the funding between equity and first debt.

General partnership: Two or more people associated legally to conduct business.

Government National Mortgage Association (GNMA):   Known as "Ginnie Mae," this government agency is a branch of HUD. Ginnie Mae guarantees principal and interest of long-term securities developed from a secondary mortgage market for government-assisted loans.

Gross income: The total receipts of an enterprise, or the total receipts excluding expenses of operating a business.

Ground rent: Rent paid by a tenant to a landlord under a ground lease.

Guaranteed loan: A loan that is guaranteed partially or fully by a specific governmental agency for the benefit of protecting a lender against possible losses.

Guaranty: A promise by one party to pay the debt of another if the party borrowing the funds fails to pay off the debt.

Hazard insurance: Insurance that protects against damage caused to property by fire, windstorm, or other common hazards. Hazard insurance is required by most lenders in an amount at least equal to the loan.

Holding costs: A term used by developers referring to the costs of owning land or property during the predevelopment stages of a project.

HUD: U.S. Department of Housing and Urban Development.

Hurdle rate: The minimum rate of return an investor will accept.

HVAC: Heating, ventilation, and air-conditioning.

Income capitalization approach: One of three methods of appraisal whereby the appraiser calculates the present value of the future stream of income from a property.

Income statement: The income statement (sometimes called a profit-loss statement) is part of the financial statement that shows the net income or net loss for a specified period of time. The income statement shows the components of revenue and expenses.

Inflation: A rise in the general price level of goods and services that decreases your purchasing power. For example, if the price of goods and services doubles, purchasing power will decrease by one half.

Institutional investors: Institutions that invest in capital assets (insurance companies, pension funds, etc.).

Institutional lender: A commercial bank, insurance company, or pension fund that provides financing for economic development projects.

Insured loan: A loan insured by a governmental agency or a private mortgage insurance company.

Interest: A charge paid for borrowed money. It is usually expressed as a certain percentage. Example, a $5,000 loan at 10% interest results in a charge of $500 for the first year for the use of this borrowed money.

Interest-only loan: A loan that requires payment of interest only rather than principal and interest. An interest-only loan typically has a shorter term than a fully amortized loan.

Interest subsidy: A grant designed to lower the interest costs of borrowing. The subsidy either goes directly to the borrower or is paid on his or her behalf to the lender.

Internal rate of return (IRR): The discount rate at which an investment has a zero net present value.

Investment syndicate: A group composed of investors, each of whom invests a sum of money and takes some share of ownership as well as a share of the risk. This enables each person to participate in investments that are larger than any one person could make alone.

Investment tax credit: A credit applied to a taxpayer's tax liability equal to a percentage of value for certain property types.

Joint venture: A legal association of two or more persons to perform a single business or service. A joint venture carries with it an expectation of profit.

Joint venture partners: The individuals or entities that associate for a joint venture.

Landbanking: The public acquisition of land and holding it in reserve for future public or private use.

Land use planning: Planning for proper use of land, taking into account such factors as transportation and location of business, industry, and housing.

Land write-down: A reduction in the price of land to below fair-market value. This land is usually sold by a public agency to help lower the costs of a redevelopment project.

Lease: A contract whereby the owner (lessor) gives the tenant (lessee) rights to possess for a stated period of time (term) for a stated amount of money (rent).

Lessee: The holder of a leasehold estate, or the tenant.

Lessor: The grantor of a leasehold estate, or the landlord.

Letter of credit: A document usually issued by a bank verifying that a borrower has credit for a specified amount.

Letter of intent: A written agreement to agree on terms and conditions of a formal contract. The letter of intent comes before the formal contract.

Leverage: The use of borrowed funds to finance a project, business, or service. A means of multiplying the availability of funds for economic development or community development programs by providing a certain amount of public funds with a proportionately larger amount of private funds.

Liabilities: Obligations of a business to pay debts such as borrowed money and merchandise purchased on credit.

Lien: A claim that someone has on the property of another as security or payment for a debt. The lien stays with the property until the debt is paid off.

Life insurance company: A company that collects premiums for life insurance and uses a portion of the premiums to provide either equity or long-term financing for large-scale economic development projects and commercial real estate projects.

Limited partnership: A partnership that restricts the liability of the limited partners to the amount of their investment.

Liquidity: The conversion of assets into cash without taking a reduction below current market value. Cash is liquid; buildings are not.

Loan terms: The provisions in a loan agreement or a deed of trust.

Local business development organization (LBDO): Nonprofit organization that specializes in business assistance such as help in collecting market data, preparing business plans, and loan applications.

Low-income housing tax credits: Tax credits associated with the investment in low-income housing. These tax credits are designed to encourage investment in low-income, multifamily housing developments.

Market: The geographical area of demand for products, goods, and/or services. This could be one neighborhood or several neighborhoods or an entire district.

Market approach to value: An appraisal term that examines the value of a specific property in a given market. The market value is defined as the most likely price that a property would command given a willing buyer and seller.

Market rent: The going rent currently charged in a market.

Market study or survey: A study of a specific area to determine its potential for supporting commercial activity. The market survey is designed to reveal information on resident shopping patterns, physical characteristics of the commercial area, and merchant business practices and to measure consumer purchasing power.

Metropolitan Statistical Area (MSA): A county or group of counties with a central city of at least 50,000 persons or an area of smaller cities clustered with a combined population of 50,000 or more.

Mortgage: A debt instrument used for real property.

Mortgage broker: A firm or an individual who arranges financing for a fee.

Mortgage constant: Percentage of the original loan balance that is represented by a constant periodic mortgage payment.

Negative cash flow: A cash loss during a period that requires an infusion of capital from investors.

Net operating income (NOI): The amount remaining after all costs, expenses, and allowances for depreciation have been deducted from the gross income of a business.

Net present value (NPV): The sum of the total present value of the annual cash flows during the holding period plus the present value of the proceeds from sale less the equity invested.

Net worth: The value of the owner's interest in the business. This is the amount by which assets exceed liabilities.

Nonrecourse loan: A loan that is secured by the property only and not by any other assets of the borrower.

One hundred percent retail location: The most important intersection in a retail district, usually where the highest volume of sales takes place.

One-stop permit system: A system set up in city government to enable businesses to arrange all city permits for business operations through one office.

Operating expense ratio: Total operating expenses divided by gross revenues.

Operating expenses: Expenses associated with a business activity, such as wages, rent, insurance, supplies, maintenance, and utilities.

Opportunity cost: The return that may have been obtained from alternative uses of capital.

Option to buy: An arrangement that permits one to buy or sell something within a specified period of time, usually according to a written agreement.

Ordinance: A rule or law established by the local governing body (city council) to control actions of citizens and the effects of their activities on others.

Overhead: Costs incurred in the sale of merchandise or services. These costs include labor, rent, utilities, insurance, office supplies, and the like.

Owner's equity: The amount by which a business's assets exceed its liabilities.

Parcel: A lot or tract of land, usually identified by ownership and a parcel number.

Partnership: A legal entity formed by two or more persons to do business. The partners must invest assets in or contribute services to the entity and must share in both the profits and losses from the business.

Pension fund: An institution that holds assets for the payment of pensions to corporate and government employees.

Points: The amount charged by the lender to the borrower at the commencement of a loan. Points are used to increase the lender's yield.

Portfolio: A group of investments held by a company or individual.

Possession: A bundle of rights held by the property owner or leasehold estate.

Present value: The current value of a future stream of income discounted by a certain rate.

Prime rate: The interest rate charged by commercial banks to business borrowers with the highest credit rating.

Principal: The amount of money borrowed that must be paid back. This amount is separate from interest and other finance charges paid for borrowing.

Proceeds from sale: The balance of funds to the seller after subtracting the loan repayment and tax liability from the net sales price (sale less broker commission).

Pro forma: A financial statement detailing current and projected revenues and expenses.

Promissory note: A written promise to pay.

Property management: The management of a real estate project or building. It includes marketing, leasing, managing, and maintenance.

Property tax: A tax imposed on real estate by municipalities.

Public works: Facilities constructed for public use and enjoyment with public funds, such as ramps, highways, and sewers, in contrast to maintenance activities, such as street cleaning and painting school buildings.

Purchase agreement: A written document in which a seller agrees to sell and a buyer agrees to buy a piece of property based on certain terms and conditions acceptable to both parties.

Quasi-public agency: Usually a nonprofit corporation with a privately appointed board of directors whose purpose is to assist governmental agencies and the private sector to more effectively improve the general living standards of our citizens. The majority of funds for such activities come from public agencies.

Rate holiday: Tax reductions given for a specified period.

Rate of return (ROR): Net operating income divided by total capital invested.

Rate of return on equity (ROE): Before-tax cash flow divided by equity investment.

Real estate investment trust (REIT):   A form of ownership of real estate that provides limited liability, no tax on entity level, and liquidity. Ownership is in the form of shares similar to common stock.

Reconciliation of value: An appraisal term that is the process of determining value using the income capitalization, market approach, and replacement cost approach to value.

Redevelopment: The physical and economic revitalization of a neighborhood or community, usually with large amounts of public funds.

Redlining: Also known as *disinvestment*, redlining is the practice by some financial institutions of designating older or declining neighborhoods or areas within a city as too risky for loans.

Refinancing: The process of paying off one loan with the money from another loan.

Regional investment corporation (RIC):   An investment company formed by a group of citizens and businesspeople to help finance specific small businesses in a specific area.

Rehabilitation: The physical improvement of an existing residential, commercial, or industrial building.

Replacement cost approach: The cost of creating a building having equal standards of materials, designs, configuration, and other improvements.

Restriction: A legal limitation in a deed on the use of property. A deed restriction may state, for example, that the owner cannot demolish the existing building without adequate replacement because the building is security for the loan.

Retail business: A business in which the owner buys products or goods from a wholesaler and sells them to customers who make personal use of what they buy.

Revolving loan fund: Usually a municipally sponsored loan program in which a specific amount of public funds is set aside to make loans for specific purposes. As loans are repaid, the funds are loaned out again.

Right-of-way: An easement on property, where the property owner gives another person the right to pass over his or her land or allow structures, such as utility poles, to be placed on the land.

Risk: The possibility of zero profits or lack of return of initial equity.

Secured loan: A loan for which collateral is pledged as security.

Service Corps of Retired Exectives (SCORE):   Management assistance for small businesses provided by retired businesspeople who are identified and registered by the Small Business Administration to provide such services.

Sinking fund: A fund to which monthly or quarterly contributions are made, usually for the purpose of replacing certain assets such as machinery or payment for a future obligation.

Small Business Administration (SBA): An agency of the state governments designed to assist small businesses with more flexible financing and less restrictive lending requirements than commercial banks.

Stabilized net operating income: Net operating income from a business, service, or property that is free from unusual items of income and expense and shows the true earning power of the entity.

Straight-line depreciation: A typical method of depreciation that takes the total depreciable amount divided by the useful life of the asset being depreciated. The resulting figure is used annually as a deduction from taxable income.

Subject property: The property being talked about or analyzed.

Subsidized housing: Residential housing constructed with financial assistance from a governmental or charitable institution, or residential housing where part of the monthly rent is paid by someone other than the tenant.

Supply and demand: An economic term that accounts for the amount of goods or services in the market being offered (supply) and being purchased or required (demand).

Tax abatement: A reduction in property taxes for a specific property over a certain period of time.

Tax basis: The original cost of ownership plus improvements minus accrued depreciation deductions.

Tax credit: A direct reduction of the tax liability.

Tax deduction: An item that will reduce the taxable income, thereby reducing the taxable liability.

Tax savings: The dollar benefit to the investor resulting from deductions or losses from a business.

Tax shelter: A deduction or credit against income. Depreciation is a tax shelter.

Taxable gain: The gain realized from the sale of an asset that is subject to either income tax or capital gains tax.

Taxable income: Income subject to tax after deductions and exemptions.

Tenant: Someone who rents from another.

Term loans: Bank loans generally made for periods from 1 to 5 years. These loans are designed to fit the particular needs of each borrower.

Time value of money: A dollar today is worth more than a dollar received at some time in the future.

Underwrite: The analysis of a loan through the examination of the creditworthiness of the borrower and the strength of the collateral.

Useful life: The period in which a business or investment asset is expected to hold economic value. It is the period over which depreciation is used.

Vacancy allowance: An amount subtracted from the gross revenues to allow for loss of income.

Value: The amount of money that can be exchanged from one product to another.

Variable-rate mortgage: A mortgage that carries with it an interest rate that can go up or down depending on the movements of rates of measurement assigned to it.

Variance: A measure of risk or the difference between the expected earnings and the actual earnings.

Venture capital: Capital subject to considerable risk. Venture capital is typically used in start-up businesses.

Warehouse: A commercial building used for the distribution or storage of goods. A warehouse may be leased or owned.

Warehousing operations: The process by which loans are held by banks or mortgage companies to sell in aggregate to the secondary market. Warehousing operations are used by commercial banks and mortgage companies.

Wrap mortgage: A new or second loan that is equal to the amount of the new loan plus the balance of the original loan.

Zoning: A local ordinance to regulate the use of private property for the benefit of the entire community. Zoning will also regulate the size of land parcels, the height of buildings, and the density of buildings. It may also regulate the floor area of buildings.

References

American Appraisal Institute. (1992). *The appraisal of real estate* (10th ed.). Chicago: Author.

Blakely, E. J. (1994). *Planning local economic development* (2nd ed.). Thousand Oaks, CA: Sage.

Blakely, E. J. (2000). *Planning local economic development* (3rd ed.). Thousand Oaks, CA: Sage.

Boston, T. D., & Ross, C. (1997). *The inner city: Urban poverty & economic development in the next century.* New Brunswick, NJ: Transaction.

Carr, J. H. (1999). Community capital and markets: A new paradigm for community reinvestment. *Neighborhood Works Journal, 17*(3), 1-4.

Porter, M. (1995, May-June). The competitive advantage of the inner city. *Harvard Business Review, 35,* 55-71.

Wurtzebach, C. H., & Miles, M. E. (1994). *Modern real estate* (5th ed.). New York: John Wiley.

Additional Resources

Bernstein, P. (1997). *Best practices of effective nonprofit organizations.* New York: Foundation Center.

Dasso, J., Shilling, J. D., & Ring, A. A. (1995). *Real estate* (12th ed.). Englewood Cliffs, NJ: Prentice Hall.

Firstenberg, P. B. (1996). *The 21st century nonprofit.* New York: Foundation Center.

Foundation Center. (2000). *The Foundation Center's guide to grantseeking on the Web.* New York: Author.

Gardner, L. (1983). *Community economic development strategies.* Berkeley, CA: Economic Development and Law Center.

Hill, E. W., & Shelly, N. A. (1990). An overview of economic development finance. In R. Bingham, E. W. Hill, & S. B. White (Eds.), *Financing economic development: An institutional response.* Newbury Park, CA: Sage.

Kaplan, R. S., & Norton, D. P. (1996). *The balanced scorecard.* Cambridge, MA: Harvard Business School Press.

Peiser, R. (1992). *Professional real estate development: The ULI guide to the business.* Washington, DC: Urban Land Institute.

Poorvu, W. J. (1993). *Real estate: A case study approach.* Englewood Cliffs, NJ: Regents/ Prentice Hall.

Shim, J. K., & Siegel, J. G. (1988). *Handbook of financial analysis, forecasting and modeling.* Englewood Cliffs, NJ: Prentice Hall.

Van Horne, J. C., & Wachowicz, J. M., Jr. (1998). *Fundamentals of financial management* (10th ed.). Englewood Cliffs, NJ: Prentice-Hall.

Index

About the Authors

Susan L. Giles is President of Giles & Company, Strategic Business Consultants and Senior Managing Director of CB Richard Ellis, Global Research and Consulting Group, where she heads the business planning practice. She is also Adjunct Professor in the master's program for the School of Policy, Planning and Development at the University of Southern California, where she teaches finance and business planning. Previously, she was a partner with Ernst & Young and Laventhol & Horwath and was President and founder of the Land Economics Group. She has been an Adjunct Professor at the Stanford University Graduate School of Business, and has taught at the University of California, Berkeley.

As a provider of business and financial consulting to the banking industry, institutional investors, and corporations, she has developed a national operational audit system and performance benchmarks for companies engaged by financial institutions and institutional investors. She has provided strategic account planning, workouts, organizational change, and financial planning to many financial institutions and technology corporations, including Regus Business Centres, Scient Corporation, Bank of America, GE Capital, California Public Employees Retirement System, California State Teachers Retirement System, Alaska Permanent Fund, Silicon Valley Bank, Sheppard Mullin Richter & Hampton, Sun Microsystems, Wells Fargo Bank, Callan Associates, Security Pacific Bank, Comerica Bank, the Bank of San Francisco, Union Bank of California, Chrysler Financial Corporation, MacPherson Oil Company, the Federal Deposit Insurance Corporation, the Resolution Trust Corporation, and others. She has worked on financial projects throughout the United States, Europe, and Mexico. She has published articles in various professional journals and is coauthor of an Urban Land Institute publication titled *Public/Private Housing Partnerships*. She holds a master's degree in city planning from the University of California, Berkeley, and a bachelor's degree in political science from the University of North Carolina, Charlotte. She also was a graduate exchange student at the Stanford University Graduate School of Business.

Edward J. Blakely is Dean of the Robert J. Milano Graduate School of Management and Urban Policy, New School University, New York City. He has held academic positions in teaching, research, academic administration, and policy development for more than 25 years. His previous positions include Lusk Professor of Planning and Development for the School of Urban Planning and Development, University of Southern California, and Professor and Chair of the Department of City and Regional Planning at the University of California, Berkeley. He is a leading scholar in the fields of planning, infrastructure, transportation, and local economic development. He is a policy advisor to the mayor of Oakland and an advisor to the Los Angeles Public School District. In addition, he serves on a number of task forces and commissions at the local, state, national, and international levels. He has served on the Board of Directors of the American Planning Association and the Nature Conservancy. He was President of the Pacific Rim Council on Urban Development (1993) and remains on the Board of Directors. He was appointed by President Clinton as Vice Chair of the Presidio Trust to serve a 2-year term (1997-1999).

Dr. Blakely is the author of four books and more than 100 scholarly articles. His book *Fortress America* (1997) with Mary Gail Snyder was named the *Choice Magazine* 1998 Academic Book of the Year. His other books are *Separate Societies: Poverty and Inequality in U.S. Cities* (1992), *Planning Local Economic Development: Theory and Practice* (Sage, 1989), and *Rural Communities in Advanced Industrial Society* (1979). He was co-recipient of the Paul Davidoff Award (1993) and a Guggenheim Fellowship (1994). He holds a joint doctorate in management and education from the University of California, Los Angeles.